WHEN YOU WERE A TADPOLE AND I WAS A FISH

WHEN YOU
WERE
A TADPOLE
AND
I WAS A FISH

AND OTHER

SPECULATIONS

ABOUT THIS AND THAT

MARTIN GARDNER

HILL AND WANG

A DIVISION OF FARRAR, STRAUS AND GIROUX

NEW YORK

Hill and Wang
A division of Farrar, Straus and Giroux
18 West 18th Street, New York 10011

Printed in the United States of America
Published in 2009 by Hill and Wang
First paperback edition, 2010

The Library of Congress has cataloged the hardcover edition as follows:
Gardner, Martin, 1914–
    When you were a tadpole and I was a fish : and other speculations
about this and that / Martin Gardner.— 1st ed.
        p.   cm.
    Includes bibliographical references and index.
    ISBN: 978-0-8090-8737-2 (hardcover : alk. paper)
        1. Science—Miscellanea.   2. Mathematics—Miscellanea.   3. Questions
and answers.   I. Title.

Q173.G355 2009
500—dc22

                                                                    2009010286

Paperback ISBN: 978-0-374-53241-3

Designed by Abby Kagan

www.fsgbooks.com

FOR JAMES RANDI,
TOP MAGICIAN,
OLD FRIEND,
AND THE WORLD'S FOREMOST
DEBUNKER OF
BOGUS SCIENCE AND CHARLATANS
WHO CLAIM
PARANORMAL POWERS

CONTENTS

PREFACE

This is another collection of articles and book reviews, of introductions to works by me and others, plus some stray pieces retrieved from obscure books of my own. The only thing these scribblings have in common is that I wrote them all.

My thanks to Dr. Sid Deutsch for proofreading galleys.

MARTIN GARDNER

Norman, Oklahoma

PART I

SCIENCE

I. ANN COULTER TAKES ON DARWIN

Ann Coulter has made a fortune by writing books that viciously insult liberals, by defending her ultra-conservative views on television talk shows, and by traveling the country giving barbed lectures. A friend recently described her with one word: cobra.

I never took Ann seriously until I read her fifth book, *Godless: The Church of Liberalism.* I wanted to find out what she had to say about evolution and intelligent design. My review of her new role as science writer first appeared in *The Skeptical Inquirer* (May/June 2008).

Ann Coulter is an attractive writer with green eyes and lopsidedly cut long blond hair, whose trademark is insulting liberals with remarks so outrageous that they make Rush Limbaugh sound like a Sunday school teacher. This is one reason why all six of her books have made *The New York Times* bestseller list and earned her fame and fortune.

Coulter's fifth book, *Godless: The Church of Liberalism,* has just been issued in paperback to provide an excuse for this review. Here are some of the book's mean, below-the-belt punches:

Monica Lewinski is a "fat Jewish girl" (p. 4).
Julia Roberts and George Clooney are "airheads" (p. 8).

Ted Kennedy is "Senator Drunkennedy" (p. 90).
The four Jersey "weeping widows" (p. 289) of men who died in
the September 11 attacks are "rabid" (p. 103), "self-obsessed"
(p. 103), and "harpies" (p. 112). "I've never seen people enjoying
their husbands' deaths so much" (p. 103).
Diplomat Joseph Wilson, whose wife was outed from the CIA,
is a "nut and liar" (p. 119) and a "pompous jerk" (p. 121). He is
likened to a "crazy aunt up in the attic" (p. 295).
Cindy Sheehan, the vocal war protester, is a "poor imbecile"
(p. 102) with an "itsy-bitsy, squeaky voice" (p. 103).
Katie Couric is a "shopworn sweetheart" (p. 295).

Liberals are repeatedly called pathetic nuts and crackpots. "[They]
are more upset when a tree is chopped down than when a child is
aborted" (p. 5). Apparently Coulter expects God to send most liberals
to hell, because she writes, "I would be crestfallen to discover any liberals in heaven" (p. 22).

Coulter has nothing good to say about any Democrat. They are all
crazy liberals who are socialists in disguise. Her latest book is titled *If
Democrats Had Any Brains They'd Be Republicans.* Here are a few other
folks who get pummeled in *Godless:*

All defenders of abortion.
All defenders of gay marriages and those who think homosexuality
is genetic.
"Hysterical" and "ugly" feminists.
Scientists who deny there could be subtle differences between the
mental abilities of men and women and between different
races.
College professors who teach students to hate God and America.
Opponents of capital punishment.
Scientists who fear global warming.
Scientists who were once afraid that AIDS would spread to
heterosexuals.

Educators who want to teach small children how to use condoms
and engage in oral and anal sex.
Opponents of nuclear power.
The staff of *The New York Times*.
Those who favor embryonic stem-cell research.
Senator John Edwards. Coulter has never apologized for
slandering him. Speaking at a political action conference
she called Edwards a "faggot" (falsely, of course). (See
Wikipedia's article on Coulter for shameful details.)
And so on.

In the last four chapters of *Godless*, Coulter suddenly morphs into
a science writer. The chapters are blistering attacks on Darwinian
evolution—the notion that life evolved gradually from simple, one-
celled forms to humans by a process that consisted of random mutations combined with the survival of the fittest. Darwin of course knew
nothing about mutations, but Coulter is concerned with modern Darwinism, which she is convinced requires some sort of superior intelligence to guide evolution.

In brief, Coulter is a dedicated believer in intelligent design, or
ID for short. Among promoters of ID, mathematician and Baptist Wil-
liam Dembski and Catholic Michael Behe are Coulter's main heroes.
Dembski, who has a degree in divinity from the Princeton Theological
Seminary, was Coulter's principal adviser on the last four chapters.

Like all IDers, nowhere does Coulter hint at how God, or a pantheistic sort of intelligence, guided evolution. There are two leading
possibilities:

1. God manipulated mutations so that new species arose, culminating finally in humans.
2. God may have allowed mutations and survival of the fittest to
   produce different breeds of a species, such as dogs and cats, but
   new species were created out of whole cloth, just as it says in the
   Book of Genesis. Like Behe and other IDers, Coulter is silent

on how God directed evolution and what sort of evidence would confirm or disconfirm the role of an intelligent designer.

This is not the place to defend in detail what Coulter likes to call the "Darwinocranks." It has been admirably done in scores of books by top scientists, all of whom Coulter considers cranks. Peter Olofson, writing tongue in cheek on "The Coulter Hoax" in the *Skeptical Inquirer* (March/April 2007), accuses Coulter of perpetrating a brilliant satire of ID rhetoric.

Let me focus instead on the transition from apelike mammals to humans. Coulter repeatedly accuses the Darwinocranks of being embarrassed by a lack of fossils that show transitional forms from one species to another. Such paucity is easily explained by the rarity of conditions for fossilization and by the fact that transitional forms can evolve rapidly. (By "rapidly" geologists mean tens of thousands of years.) Moreover, transitional fossils keep piling up as the search for them continues.

Nowhere are transitional forms more abundant than in the fossils of early human skeletons and the skeletons of their apelike ancestors. Consider the hundreds of fossils of Neanderthals. H. G. Wells, in a forgotten little book titled *Mr. Belloc Objects* (see chapter 4 of this book), defends evolution against ignorant attacks by the Catholic writer Hilaire Belloc. In chapter 4 of his book, Wells has this to say about Neanderthals:

> When I heard that Mr. Belloc was going to explain and answer the *Outline of History*, my thought went at once to this creature. What would Mr. Belloc say of it? Would he put it before or after the Fall? Would he correct its anatomy by wonderful new science out of his safe? Would he treat it like a brother and say it held by the most exalted monotheism, or treat it as a monster made to mislead wicked men? He says nothing! He just walks away whenever it comes near him.
>
> But I am sure it does not leave him. In the night, if not by day, it must be asking him: "Have I a soul to save, Mr. Belloc? Is that Heidelberg jawbone one of us, Mr. Belloc, or not? You've forgotten me, Mr.

Belloc. For four-fifths of the Paleolithic age I was 'man.' There was no other. I shamble and I cannot walk erect and look up at heaven as you do, Mr. Belloc, but dare you cast me to the dogs?"

No reply.

Coulter is as silent as Mr. Belloc about Neanderthals and about the even earlier, more apelike skeletons. I doubt if they trouble her sleep; I doubt if anything troubles Coulter's sleep. Does she think there was a slow, incremental transition from apelike creatures to Cro-Magnons and other humans? Or does she believe there was a first pair of humans?

Let's assume there was a first pair. Does Coulter think God created Adam out of the dust of the earth, as Genesis describes, then fabricated Eve from one of Adam's ribs? Or does she accept the fact that the first humans were the outcome of slow, small changes over many centuries? If the transition was sudden, then Adam and Eve were raised and suckled by a mother who was a soulless beast!

This is a bothersome dilemma for all Christians who believe in the crossing of a sharp line from beast to human. It is a dilemma about which I once wrote a short story called "The Horrible Horns." If interested, you can find it in my book *The No-Sided Professor and Other Tales of Fantasy, Humor, Mystery, and Philosophy.*

We know from a footnote on page 3 of *Godless* that Coulter considers herself a Christian. But what sort of Christian? The word has become enormously vague. Today one can call oneself a Christian and hold beliefs that range from the fundamentalism of Jerry Falwell and Billy Graham, through the liberal views of mainline Protestant ministers and Catholic liberals such as Hans Kung and Gary Wills, to the atheism of Paul Tillich. Tillich did not believe in a personal God or an afterlife, two of the central doctrines of Christ's teachings, yet he is considered by many Protestants to be one of the world's greatest Christian theologians!

Wikipedia's article on Coulter quotes her as saying, "Christ died for my sins . . . Christianity fuels everything I write." This sounds like something an evangelical Protestant would say. On the other hand, in

Godless Coulter quotes a remark by G. K. Chesterton (p. 10), who is almost never quoted today except by Catholics. Is Coulter a Protestant or a Catholic? Or some other kind of Christian?

Although I am not a Catholic, allow me to cite a famous passage from Chesterton's introduction to his book *Heretics*:

> But there are some people, nevertheless—and I am one of them— who think that the most practical and important thing about a man is still his view of the universe. We think that for a landlady considering a lodger, it is important to know his income, but still more important to know his philosophy. We think that for a general about to fight an enemy, it is important to know the enemy's numbers, but still more important to know the enemy's philosophy. We think the question is not whether the theory of the cosmos affects matters, but whether, in the long run, anything else affects them.

Coulter, you are merciless in bashing liberals and atheists, so *please* let us know what church you attend. It would clear the air and shed light on your peculiar personality and on the background for all your insults, especially your blasts at Darwinians.

Here's another simple question to ponder: Why do you suppose God provided men with nipples?

2. ISAAC NEWTON'S VAST OCEAN OF TRUTH

There are two reasons for viewing Newton's career as awesome: his stupendous discoveries in mathematics and physics, and the equally stupendous stupidity of his theology. My review of Peter Ackroyd's *Newton* (2008) appeared in *The New Criterion* (April 2008). For more of my opinions about Newton, see my article "Isaac Newton: Alchemist and Fundamentalist" in *The Skeptical Inquirer* (September/October 1996), reprinted in *Did Adam and Eve Have Navels?* (New York: Norton, 2000).

Peter Ackroyd is an acclaimed and prolific British novelist, poet, dramatist, and biographer. He has written biographies of William Shakespeare, Charles Dickens, William Blake, Thomas More, Oscar Wilde, Edgar Allan Poe, Ezra Pound, and T. S. Eliot. His history of London was a bestseller. His new biography of Isaac Newton follows the lives of Geoffrey Chaucer and J.M.W. Turner.

There has been a raft of recent biographies of Newton, notably Richard Westfall's *Never at Rest.* Why another one? The answer is that a brief life of Newton meets a widespread need. Longer biographies may tell you more about a person than you care to know. Although there are no new or startling revelations in Ackroyd's book, its facts are accurate, his judgments sound, and his writing a great pleasure to read.

Isaac Newton (1643–1727) was a strange, improbable blend of a great mathematician and physicist, one of history's greatest, with the mind-set of an ignorant, naive fundamentalist. A practicing Anglican, he never doubted that God created the entire universe in six literal days, that He once drowned every human and beast except for Noah and his companions, that Eve was fabricated from Adam's rib, that Lot's wife turned into a pillar of salt, that Moses parted the Red Sea, and that the prophecies of Daniel and the Book of Revelation came straight from the Almighty and are certain to be fulfilled.

Newton tried to calculate the exact date of Jesus's return to earth. He set a precise year for the creation that was half a century *later* than Bishop Ussher's famous 4004 B.C.E. He was convinced that the Catholic Church was the Antichrist of the Book of Revelation. After his death, Ackroyd tells us, Newton left a manuscript on biblical prophecy that ran to 850 pages.

Newton's single great departure from Anglican orthodoxy was his opposition to the Trinity. Jesus was indeed the Son of God, but he was not Jehovah. "We should not pray to two Gods," Newton wrote. The notion that Jesus was God in human flesh was a heresy perpetuated by Rome. Newton carefully concealed his anti-trinitarianism to avoid being expelled from Cambridge University where for decades he was a professor, ironically at Trinity College.

Another aspect of Newton's curious, complicated life was his obsession with alchemy. He owned and studied all the books on alchemy he could obtain, and spent endless days in his laboratory trying vainly to turn base metals into gold. His unpublished writings on alchemy, though smaller than his writings on Bible prophecy, ran to more than a million words, far exceeding everything he wrote about physics and astronomy. Ackroyd cites John Maynard Keynes's celebrated Cambridge lecture on Newton's secret records about alchemy. He found nothing of the slightest value to science.

Late in life, Newton suffered a mental breakdown that lasted more than a year. It has been suggested it was the result of poisoning by the mercury he used in his alchemical experiments. Others believe Newton was the victim of a bipolar disorder that triggered a deep depression.

It is hard now to comprehend, but only a small fraction of Newton's long life was devoted to investigating God's laws of nature. In a few years of his mid-twenties he invented calculus, found that white light was a mixture of all colors, explained for the first time the rainbow, and constructed one of the earliest reflecting telescopes. His greatest discovery, of course, was that gravity, which holds us to the earth and makes apples fall, is the same force that guides the path of our moon, our sister planets, and our comets. What else might he have discovered had he not squandered his energy and talents on alchemy and Biblical exegesis!

Gravity, Newton wrongly believed, acts instantaneously at a distance. Its nature remained a total mystery. Newton knew its force varied directly with the product of two masses, and inversely with the square of the distance between them, but its cause, he said, "I do not pretend to understand." Not until Einstein was it partly explained by the curvature of space-time.

Light, Newton believed, was corpuscular, composed of minute particles. In this he was half right. Today, light is known to be both a particle and a wave.

Ackroyd is good in describing Newton's complex personality, which is almost as bizarre as his beliefs. In Ackroyd's words he was "suspicious and secretive," with "a great capacity for anger and aggression." There are records of him smiling, only one of him laughing. It occurred when someone asked him what is the value of studying Euclid.

In his younger years, Newton often slept in his clothing. Even when not absorbed by work he would go without eating or eat standing up. He never exercised or had any hobbies. Going out, he frequently forgot to comb his hair or fasten his stockings. In his elderly years, when he was Warden of the Royal Mint, he was ruthless in seeing that counterfeiters were hanged.

Newton had almost no interest in art, music, literature, or women. Here is his account of his only attendance at an opera. "The first act I heard with pleasure, the second stretched my patience, at the third I ran away." He once dismissed poetry as "ingenious fiddle-faddle."

During his depression Newton wrote a curious letter to the philosopher John Locke:

> Sir, being of opinion that you endeavored to embroil me with woemen [an odd if for him appropriate spelling] & by other means I was so much affected with it as that when one told me you were sickly & would not live I answered twere better if you were dead. I desire you to forgive me this uncharitableness.

Puzzled, Locke replied that of course he was forgiven, and they remained friends.

Newton was always proud of his discoveries and furious when others claimed to have made similar findings earlier. His most bitter quarrel was with the German philosopher and mathematician Leibniz over the discovery of calculus. Leibniz was unquestionably the first to publish, and his notation proved superior to Newton's. Today's opinion is that each made his discovery of calculus independently, not knowing of the other's work.

There has been speculation that Newton was gay. Ackroyd finds the evidence slim. It rests on nothing more than Newton's lack of interest in women, and his friendship in later life with a much younger Swiss mathematician who idolized him. At one time, the two hoped to share lodgings, but nothing came of the plan.

If Newton were to return to earth today, he would of course be overwhelmed by developments in physics and astronomy. He would be less astounded, I suspect, by cars, trains, airplanes, even electric lights, but a desk calculator and a television screen would seem sheer sorcery to him. As for the occasional scientist who believes that physics is on the verge of discovering everything, he would surely have only contempt.

There is a famous passage in which Newton suggests how little science knows, perhaps how little it can know. The passage is often quoted, but deserves repeating over and over again. Here is how Ackroyd gives it in his last chapter but one:

> I don't know what I may seem to the world, but as to myself, I seem to have been only like a boy playing on the sea shore, and diverting myself in now and then finding a smoother pebble or a prettier shell than ordinary, whilst the great ocean of truth lay all undiscovered before me.

3. BULL'S-EYES AND FAMOUS FUMBLES

For almost ten years I wrote a puzzle column in *Isaac Asimov's Science Fiction Magazine*. The following chapter reprints a December 1985 column in which I reported on some outstanding instances of amazingly accurate predictions about the future of science, and other cases of predictions hopelessly wide of the mark.

Considering the fact that thousands of predictions are made every year in science fiction (SF) stories around the globe, it is not surprising that there are occasional hits of startling accuracy, like shooting a shotgun at a target. Of course there are even more whopping misses. Sometimes hits and misses accompany one another. Jules Verne scored a fantastic hit when he had the first spaceship shot around the moon from a spot in Florida, but his ship was blasted off by a gigantic underground cannon. Hundreds of SF tales anticipated the moon walks. As far as I know, only one guessed that the first walk would be watched on earth's television screens: *Prelude to Space*, a novel by Arthur C. Clarke. It was first published in *Galaxy* (February 1951), and later reprinted under other titles.

It would be no small task to draw up a complete list of H. G. Wells's hits and misses. His most spectacular success was in the chapter that opens *The World Set Free* (1914), in which Wells tells how the atom was first split. The novel has a second world war starting in the forties, and there is a graphic description of an "atomic bomb" (yes, Wells used the term!) dropped on the enemy. But the bomb is held by a person who drops it through an opening in the bottom of a plane.

In his 1902 collection of prophetic essays, *Anticipations*, Wells correctly foresaw wide asphalt thruways, looping over and under at intersections, and with a dividing barrier between traffic running in opposite directions. A chapter on twentieth-century warfare is amazingly accurate in many ways, but the air battles are fought by men in balloons, and Wells had this to say about submarines: "I must confess that my imagination, in spite even of spurring, refuses to see any sort of submarine doing anything but suffocate its crew and founder at sea." It has been justly said that Wells hit the mark more often in his SF than in his nonfiction. In *Social Forces in England and America*, published the same year as *The World Set Free*, he speaks of "the tapping of atomic energy, but I give two hundred years before that."

For decades I have been trying to gather a complete run of Hugo Gernsback's marvelous *Science and Invention*, especially during its golden age, the 1920s. The magazine's lurid covers are an amusing mix of hits and misses. Among the hits: helicopters carrying girders for skyscraper construction, the use of flamethrowers in warfare, and (my favorite) a man and woman embracing, with wires attached to various parts of their bodies to measure heartbeat, respiration, perspiration, and so on. It illustrated Gernsback's article on the scientific study of sex. Among the misses: a giant robot policeman, and a picture of what a Martian would look like. Gernsback's *Ralph 124C 41+* is probably the worst SF novel ever published (it ends with the awful pun "one to foresee for one"), yet it also contains some of the most accurate predictions ever made in such a novel.

If you're interested in outlandish misses about the future, I recommend *The Experts Speak: The Definitive Compendium of Authoritative Misinformation*, by Christopher Cerf and Victor Navasky. On hits by SF writers, see the article "Prediction" in *The Science Fiction Encyclopedia*, edited by Peter Nicholls. Listed below, from my collection, are some outstanding instances of astonishing anticipations by writers outside the SF field. See if you can guess some of the authors and the centuries in which they wrote.

1. "Clothes hung up on a shore which waves break upon become moist, and then get dry if spread out in the sun. Yet it has not been seen in what way the moisture of water has sunk into them nor again in what way this has been dispelled by heat. The moisture therefore is dispersed into small particles which the eyes are quite unable to see."

2. "The primary elements of matter are, in my opinion, perfectly indivisible and nonextended points; they are so scattered in an immense vacuum that every two of them are separated from one another by a definite interval."

3. "[T]wo lesser stars, or satellites, which revolved about Mars, whereof the innermost is distant from the centre of the primary planet exactly three of the diameters, and the outermost five; the former revolves in the space of ten hours, and the latter in twenty-one and an half."

4. "If Mr. B will drink a great deal of water, the acrimony that corrodes his bowels will be diluted, if the cause be only acrimony; but I suspect dysenteries to be produced by animalculae which I know not how to kill."

5. "I know a way by which 'tis easy enough to hear one speak through a wall a yard thick . . . I can assure the reader that I have by the help of distended wire propagated the sound to a very considerable distance in an instant, or with as seemingly quick a motion as that of light . . . and this not only in a straight line, or direct, but in one bended in many angles."

6. "Such changes in the superficial parts of the globe seemed to me unlikely to happen if the Earth were solid to the centre. I therefore imagined that the internal parts might be a fluid more dense, and of greater specific gravity than any of the solids we are acquainted with; which therefore might swim in or upon that fluid. Thus the surface of the globe would be a shell, ca-

pable of being broken and disordered by the violent movements of the fluid on which it rested."

7. "Would it be too bold to imagine that, in the great length of time since the world began to exist, perhaps millions of ages before the commencement of the history of mankind—would it be too bold to imagine that all warm-blooded animals have arisen from one *living filament*, which the great First Cause imdued with animality, with the power of acquiring new parts, attended with new propensities, directed by irritations, sensations, volitions, and associations, and thus possessing the faculty of continuing to improve by its own inherent activity and of delivering down these improvements by generation to its posterity, world without end?"

8. *For I dipt into the future, far as human eye could see,*
   Saw the Vision of the world, and all the wonders that would be;

   Saw the heavens fill with commerce, argosies of magic sails,
   Pilots of the purple twilight, dropping down with costly bales;

   Heard the heavens fill with shouting, and there rain'd a ghastly dew
   From the nation's airy navies grappling in the central blue;

   Far along the world-wide whisper of the south-wind rushing warm,
   With the standards of the peoples plunging thro' the thunderstorm;

   Till the war-drum throbb'd no longer, and the battle-flags were furl'd
   In the Parliament of man, the Federation of the world.

ANSWERS

1. Lucretius, *De Rerum Natura* (On the Nature of Things), circa 99 B.C.E. This accurate description of evaporation shows that ancient Greek and Roman particle theory had more empirical support than some historians of science like to admit.

2. Roger Joseph Boscovich, *Theoria Philosophiae Naturalis* (Theory of Natural Philosophy), 1758. In today's particle theory, matter

is believed to be made of six kinds of leptons and six kinds of quarks, all pointlike, with no internal structure.

3. Jonathan Swift, "A Voyage to Laputa," in *Gulliver's Travels*, 1726. Mars's two moons were not discovered until 1877. Phobos, the innermost moon, revolves in a trifle more than seven hours, and Deimos, the outermost moon, in about thirty-one hours. That Mars had two moons had earlier been predicted by Kepler. This was probably the basis of Swift's account.

4. Samuel Johnson, in a letter to Mrs. Thrale, November 12, 1791.

5. Robert Hooke, British physicist, in *Micographia*, 1664.

6. Benjamin Franklin, in a letter to Abbé Soulavie, September 22, 1782.

7. Erasmus Darwin, Darwin's grandfather, in *Zoonomia*, 1794.

8. Alfred Tennyson, *Locksley Hall*, 1886.

Now see if you can name the single scientist responsible for the following pratfalls:

"The talking motion picture will not supplant the regular silent motion picture . . . There is such a tremendous investment to pantomime pictures that it would be absurd to disturb it." (1913)

"It is apparent to me that the possibilities of the aeroplane, which two or three years ago was thought to hold the solution to the [flying machine] problem, have been exhausted, and that we must turn elsewhere." (1895)

"In fifteen years, more electricity will be sold for electric vehicles than for light." (1910)

"There is no plea which will justify the use of high-tension and alternating currents, whether in a scientific or a commercial sense . . . My personal desire would be to prohibit entirely the use of alternating currents. They are unnecessary as they are dangerous."* (1889)

*All the remarks were made by Thomas Edison. I found the first three in *The Experts Speak* (cited earlier), and the fourth in a section on predictions in *A Random Walk in Science*, compiled by R. L. Weber (1973).

4. MR. BELLOC OBJECTS

Today few admirers of H. G. Wells and of the Catholic writer and historian Hilaire Belloc are aware that the two men once engaged in a bitter dispute over evolution. It began when Belloc attacked Wells's *Outline of History* in a series of articles later reprinted as a book titled *A Companion to Wells's Outline of History*. An angry Wells retaliated with a now rare little book, *Mr. Belloc Objects*. It prompted Belloc to fire back with another rare book, *Mr. Belloc Still Objects*.

The Battered Silicon Dispatch Box, in Canada, has brought out both books in a single volume. What follows here is my introduction to this book.

Wells's monograph is now dated in several respects. For instance, Wells had no way to know that the famous Piltdown skull was a forgery. However, *Mr. Belloc Objects* is still an excellent, very funny introduction to Darwinian evolution. Belloc's feeble reply is a classic example of the inability of so many Christians to comprehend how evolution could be God's method of creation. This is a controversy not soon to fade.

H. G. Wells's rare little book, *Mr. Belloc Objects*, is almost totally forgotten today except by Wells collectors. In 1926, when this book was first printed, Wells had hit the jackpot with his *Outline of History*. It had become a worldwide bestseller, and had earned for Wells a considerable fortune.

Histories of the world had been written earlier, but Wells's *Outline* differed from them in two ways: its range was wider, and it opened with a history of prehistoric humanity that assumed the soundness of Darwinian evolution. Hilaire Belloc, a well-known British writer and an ultra-conservative Catholic, wrote a series of articles attacking Wells's *Outline*, especially his defense of Darwin. Belloc's articles were widely published in Catholic periodicals, and later issued as a book titled *A Companion to Mr. Wells's Outline of History.*

Wells was furious. Not just because evolution was dismissed as bad science, but because Belloc was savage in his personal attacks on Wells. It was the first and last time that Wells was goaded into replying in kind to ad hominem criticism. His little book is not only very funny but one of the strongest pieces of rhetoric ever written in defense of evolution against a form of creationism known today as Intelligent Design (ID).

The creationism Belloc defends was not, of course, the crude fundamentalism of Protestant young earthers convinced that God created the entire universe in six literal twenty-four-hour days. Belloc was not a young earther. He accepted evolution in a sense, but insisted that each species was a new creation. In just what way it was a new creation Belloc does not say. Like today's IDers, Belloc leaves this detail hopelessly vague. Did God create out of whole cloth the first pigs? Or did he merely guide mutations in such a way that pigs were suddenly born to non-pig parents?

Today's IDers are equally mute on this point. As attorney Phillip Johnson says over and over again in his books defending ID, it is not necessary for opponents of Darwin to explain exactly how God guided evolution. It is only necessary to make clear the inadequacy of explaining the origin of species by random mutations followed by survival of the fittest.

The question becomes especially bothersome with respect to the origin of humans. As Wells so beautifully argues in his last chapter, proponents of ID are haunted by the fossils of Neanderthals. Were they true humans, with immortal souls, or were they merely higher apes?

Exactly how did this monumental transition occur? Surely Belloc did not believe God created Adam out of the dust of the earth, then fabricated Eve from one of Adam's ribs. But if the Genesis account is mythological, exactly how was the transition made? Were the first humans reared and suckled by a mother who was a beast? Belloc is as silent on this question as are today's IDers.

At the time Belloc lambasted Wells, the Catholic Church by and large rejected all forms of evolution. Consider the tragic story of St. George Jackson Mivart (1827–1904). He was a much-admired British zoologist, author of many books including a massive volume titled *The Cat*, and a student of the great Thomas Huxley. He was also a devout Catholic, but a Catholic with extreme liberal views. Over and over again, in scholarly journals, and in his book *On the Genesis of Species*, Mivart defended the evolution of all forms of life, including humans, but by a process governed by God and with the proviso that immortal souls were infused into the first men and women. Repeatedly he warned his Church that by opposing evolution it was making a foolish mistake similar to the mistake it had made in persecuting Galileo for claiming the earth went around the sun. In brief, Mivart held views on evolution identical with the views of almost all of today's Catholic philosophers and theologians. Poor Mivart! He was too far ahead of his time. The Church branded him a heretic, excommunicated him, and denied him a Christian burial.*

Today almost all liberal Christians, Protestant or Catholic, regard evolution as God's method of creation, no more requiring miracles along the way than the origin of the solar system required frequent pushes by the Creator, as Newton believed, to keep the planets in orbit.

Almost all of today's IDers are conservative Christians. Phillip Johnson is an evangelical Presbyterian. David Berlinski is a conservative Baptist. Michael Behe, the most persuasive of the lot, is a Catholic. Berlinski recently teamed up with Ann Coulter to help her write the part of her book *Godless* in which she bashes Darwin.

---

*On George Mivart, see chapter 9 of my *On the Wild Side* (Amherst, NY: Prometheus, 1992), and Jacob Gruber's excellent biography, *A Conscience in Conflict* (New York: Columbia University Press, 1960).

How did Belloc react to Wells's attack? He promptly wrote a book titled *Mr. Belloc Still Objects.*

How do things now stand with the Roman Church? I am pleased to report that Pope John Paul II declared evolution to be more than a theory and worthy to be taught in all Catholic schools. The Church is moving slowly and cautiously in a liberal direction—that is, in the direction promoted by such Catholic thinkers as Hans Kung, in Germany, and in the United States by Gary Wills, Father Andrew Greeley, and many others. Hopefully it will continue to glide in that direction. If it goes the other way, it will be a sad day for both the Church and the world.

5. MR. BELLOC STILL OBJECTS

After the text of *Mr. Belloc Still Objects* in the Battered Silicon edition of Belloc's and Wells's books, I added the brief epilogue to Belloc's book reprinted here.

Belloc learned nothing from Wells's attack. His rebuttal continues the same insults, the same below-the-belt punches. As in his original articles on Wells's *Outline of History*, it is not so much what Belloc says that reveals his ignorance of biology and geology, it is what he doesn't say.

What he doesn't say is how he thinks the "fixed types" came into being. Did God somehow manipulate the genetic information in sperm and eggs so that mothers gave birth to widely different life forms, or did the Almighty create the first fixed types from nothing, or from, as in Adam's case, the earth's dust? IDers today are similarly tongue-tied on this fundamental question.

Belloc seems unable to comprehend that types *seem* fixed because they are end points on branches of the evolutionary tree. Intermediate forms are all in the past. Fossilization is a rare event. We get only glimpses of extinct missing links. Belloc seems to think that because a horse can't mate with a cow, the way lions can mate with tigers, horses were always horses and cows were always cows.

I have assumed that Belloc was not a Catholic fundamentalist who took the two Genesis accounts of creation to be literal history, and that he understood the word *day* of course to mean a period of many millions, or even billions, of years. Now I'm not so sure. I was startled by a sentence on page 88 of *Mr. Belloc Still Objects* in which Belloc says the Fall of Man took place 5930 years ago "in the neighborhood of Baghdad." How Belloc arrived at that date and place beats me. Wells must have guffawed when he read this! Belloc certainly believed the Fall was some sort of event, such as Adam and Eve chewing a forbidden fruit from the tree of knowledge of good and evil. It is hard to accept that Belloc could believe this, including the role of a talking snake, but perhaps he did. If not, just what *was* the Fall?

Belloc generously grants that Wells is right on a few trivial points, and he promises to make corrections in his articles when they soon go into a book. Some of Belloc's criticisms of Wells are right on target. That Wells was an atheist is undeniable, though this was not always the case. In his younger years he believed in a limited or finite God, a God growing in time as portrayed in the philosophies of such "process theologians" as Samuel Alexander and Charles Hartshorne. Wells defended such a deity in his war novel *Mr. Britling Sees It Through*, and in a nonfiction work titled *God the Invisible King*. Later he decided he was really an atheist.

Belloc is also right in accusing Wells of having a low opinion of the Roman Church. He once likened the Church to a huge dinosaur roaming the earth and refusing to become extinct. Shortly before he died he wrote *Crux Ansata*, a savage indictment of Roman Catholicism. Belloc even catches Wells in a whopping error, one often made by non-Catholics, when Wells confuses the Virgin Birth of Jesus with Mary's Immaculate Conception!

What about Belloc's many quotations from scientists expressing doubts about Darwinism? My guess is that most of the men quoted were finding fault with Darwin's views about the *process* of evolution, not the fact. Darwin, it must be recalled, knew nothing about mutations. He was a follower of Jean-Baptiste Lamarck, a French biologist who believed that an animal's striving to improve its chances of survival caused favorable genetic changes. For example, a giraffe's efforts to eat leaves high on trees would somehow be transmitted to sperm and eggs with the result that its descendants' necks would become a trifle longer. Similarly, disuse of an organ would cause it to weaken and perhaps disappear. In Belloc's day Lamarckism was being abandoned for lack of evidence. Today the inheritance of acquired traits has been totally discredited.

Darwinism of course is a fuzzy term. In a narrow sense, Darwin's Lamarckism has been so thoroughly discarded that one can say Darwinism indeed expired. But in a broader sense, Darwinism is alive and well. It is a fact, accepted today by all biologists and geologists, with very few exceptions, that life began on earth with one-celled forms, then slowly altered by random mutations which were either beneficial or harmful, or neither. Favorable mutations have tended to survive. Harmful mutations tended to die off. I suspect that most, perhaps not all, the scientists quoted by Belloc actually were firm believers in evolution but had differences with Darwin over the exact way in which evolution took place.

Consider the remark by William Bateson, a famous British geneticist. Bateson doesn't deny evolution. He merely expresses his belief that factors other than natural selection played a dominant role in the process. In Bateson's time there was considerable controversy over the mechanisms of evolution, controversy that persists today. But Bateson never doubted that all life could be graphed by a single tree.

Hans Driesch, a German biologist, was another firm believer in evolution who differed from Darwin only on technical matters. Eberhard Dennert, Belloc's next expert, was an obvious crank. His 1904 book *The Deathbed of Darwinism* is described by Stephen Jay Gould in his book *The Panda's Thumb* as ranking "right up there with bumblebees can't fly" and "rockets won't work in a vacuum." Dennert predicted evolution's total demise by 1910!

Belloc annoyingly never gives first names to his experts, so I have been unable to learn anything about a man with the last name of Dwight. He sounds like a simpleminded creationist. There are lots of scientists with the last name of Morgan. I don't know which one Belloc is quoting, but his disagreement with Darwin is clearly over the process, not the fact, of evolution. We are told nothing about a 1909 book by, I assume, F. le Danrée, a French biologist, except that it is a "crushing blow" against evolution.

Sitting in my apartment, in an assisted living facility (I'm ninety-four), with only a computer on hand as a research tool, I have been unable to learn anything about Edward Dawson Cope, an American paleontologist who is probably the Cope Belloc has in mind. Nor could I find much about Yves Delarge, a French zoologist, or Karl Wilhelm Nägeli, a Swiss botanist. I've drawn a total blank on any scientist named Korchinsky. Perhaps someone more familiar than I with the early history of evolution can tell us just what the men quoted by Belloc actually believed.

Even today scientists wrangle over the mechanisms of evolution. Gould, for instance, promoted a controversial view called "punctuated equilibrium." His theory emphasizes the fact that some species, trilobites for example, remained unaltered for millions of years, while others changed radically over a period of a few thousand years. Gould was of course a thoroughgoing Darwinian, yet his opinions have been strongly opposed today by Richard Dawkins and others.

As far as I know, Wells made no effort to reply to Belloc's *Still Objects* book. Regardless of his objections, Belloc must still have been haunted by the Neanderthals. Were they apelike humans or human-like apes? Was there a turning point in history, according to Belloc some six thousand years ago, when the first humans suddenly appeared on old earth like a magician's beautiful assistant stepping out of a previously empty container? Maybe some expert on Belloc's opinions can tell us what Belloc fails to reveal in his forgotten little volume.

PART II

BOGUS SCIENCE

6. WHY I AM NOT A PARANORMALIST

When Carl Sagan, at a meeting of the American Association for the Advancement of Science, attacked the wild, ignorant opinions of Immanuel Velikovsky, he was lambasted by his colleagues! Astronomers had more important tasks, they argued, than to waste time debunking outlandish claims. I do not think Sagan was wasting anyone's time. Nor did he waste time later when he wrote entire books about pseudoscience.

Our nation is weakened when large numbers of citizens, including members of Congress and even presidents, are scientific illiterates. We recently had a president and first lady, Ronald Reagan and his wife, Nancy, who passionately believed in astrology! And former President George W. Bush is on record as favoring the teaching of creationism in public schools alongside what he regards as Darwin's dubious "theory" of evolution!

I have spent a good part of my writing career debunking bogus science. In the following chapter, excerpted from my confessional *The Whys of a Philosophical Scrivener*, I do my best to explain why I am a skeptic of paranormal claims made on behalf of psi, or parapsychological psychic phenomena or powers.

No conceivable event, however extraordinary, is impossible; and, therefore, if by the term miracles we mean only "extremely wonderful events," there can be no just ground for denying the possibility of their occurrence.

—Thomas Henry Huxley, *Hume*

Science is a search for reliable knowledge about the world: how the universe (including living things) is structured and how it operates. The search began when primitive minds first became aware of such obvious regularities as day and night, the beating of hearts, the falling of dropped objects. Slowly over the millennia the search has been refined and systematized, first by better observations, then by making experiments and by inventing theories and new observing instruments.

In a rough way scientific information can be divided into three parts: facts, laws, and theories. No sharp lines separate these groupings, but if we did not give names to parts of continua we would be unable to talk. *Day* and *night* are useful terms in spite of—or, rather, because of—their fuzzy boundaries. If a woman tells you she dreamed about you last night, you are not likely to respond with, "Exactly what do you mean by *night*?"

For a typical fact let's take the statement *Mars has two moons*. A typical law: planets travel in elliptical orbits. A typical theory: general relativity. The first is a fact because it refers to a unique case. The second is a law because it generalizes from particular cases to all planets that go around single suns. The third is a theory because it deals with unobservable entities such as fields, and because it explains both facts and laws.

In all three areas—facts, laws, theories—scientific knowledge is never certain. Mars may have a third moon so small it hasn't yet been detected. Laws may be only crude first approximations. Theories may turn out to be mistaken. No statements about the world are absolutely certain, though some can be assigned a degree of credibility that everyone agrees is practically indistinguishable from certainty. Who would deny that elephants have trunks or that you who read this are not a cockroach?

Only in pure logic and mathematics can statements be deemed absolutely certain, but for this kind of truth a stupendous price is paid. The price is that such statements say nothing about the world. To apply abstract mathematics to physical things we have to make all sorts of assumptions, which Rudolf Carnap called correspondence rules. Why do two elephants plus two elephants make four elephants? It is because we have strong empirical grounds for the correspondence rule that says elephants combine like positive integers. But two drops of water plus two more drops can make one big drop. Applied mathematics is never absolutely certain because there is always the possibility, however remote, that a mathematical model is not absolutely accurate.

It does not follow from the vagueness of scientific method and the controversies surrounding it that science does not genuinely advance. Nor does it follow that there is no rational basis for evaluating competing theories. Everyone agrees that science moves forward by a constant testing of new hypotheses, most of which have to be discarded. The process has been compared to that of organic evolution. Genetic mutations are copying errors that almost always lower the probability of a species' survival. Organisms with unfavorable mutations tend to die out; those with the rare favorable mutations tend to survive. In a similar way, the continual proposals of unorthodox theories, most of which turn out to be flawed, are essential to the progress of science. Establishment journals, contrary to what some people think, are crammed with just such offbeat speculations, and the surest road to fame is to advance a crazy theory that is eventually confirmed, often after intense resistance by skeptics. Such resistance is both understandable and necessary. Science would be total chaos if experts quickly embraced or even tried to disconfirm every eccentric theory that came along.[1]

Why do I stress such pedestrian views? Because, as a skeptic of paranormal claims, I am perpetually accused by parapsychologists of stubbornly denying the very possibility that psi forces such as ESP (telepathy, clairvoyance, precognition) and PK (psychokinesis) exist. It is a foolish accusation. No skeptic of psi known to me rules out the possibility of psi. Any law or theory is possible that is not logically inconsistent. It is possible that Wilhelm Reich discovered a hitherto unrecognized force, orgone energy. It is possible, as Scientologists contend, that a day-old human embryo records all words spoken to or by its mother. It is possible that you and I lived previous lives. It is possible that the sun stood still when Joshua commanded it, that Jesus walked on the water, turned water into wine, and raised people from the dead. It is possible that God made Eve from Adam's rib. It is possible that the earth is a hollow globe and we are living on the inside surface. All statements about the real world have varying degrees of credibility, with one and zero as unreachable limits, even though we can get within an infinitesimal distance from either end.

It is obvious, at least to me, that science has only begun to unravel the mysteries of the universe. If by *paranormal* you mean all the laws and theories not yet discovered, then all scientists believe in the paranormal. If we could somehow obtain a physics textbook of the twenty-fifth century, it is a good bet that it would contain information that no physicist now alive could guess or even understand. Every branch of science has frontiers where it pushes outward into vast unexplored jungles that swarm with paranormal surprises. In this sense of *paranormal* I am a paranormalist of the most extreme sort. I firmly believe there are truths about existence as far beyond our minds as our present knowledge of nature is beyond the mind of a fish.

When I say I am not a paranormalist I use the word in the way it is used today in ordinary discourse. We are living at a time of rising interest, on the part of an uninformed public, in wild beliefs that the entire science community considers close to zero in credibility. It is this vaguely defined body of crazy science that writers and television commentators mean when they use the term *paranormal*.

Need I summarize some of the things you are likely to see when a pseudodocumentary film explores the paranormal? UFOs bearing aliens from outer space; mysterious deadly forces in the Bermuda triangle; the power of the Great Pyramid's shape to preserve food and sharpen razor blades; the reality of demon possession, psychic surgery, dowsing, astrology, palmistry, numerology, biorhythms; the claims of Transcendental Meditators that one can learn to levitate, become invisible, and walk through walls; the human aura; out-of-body travel; poltergeists; Spiritualism; phone calls from the dead; statues of Mary that weep and bleed; the sensitivity of plants to thoughts; and (on a more credible level) ESP and PK.

In a competitive society such as ours the owners of the print and electronic media always pander to public taste, and there is much to be said for such pandering. If large numbers of people relish bad art, music, and literature, bad movies and television shows, and bogus science, who can say they should not be given the opportunity to read, hear, and see what they like? Any attempt at censorship, even of the most blatant pornography, is always a risky business in an open society. No one who values liberty and democracy wants to see legislation on any government level, except in extreme cases, that tells the media what they cannot do.

There are, however (at least I so believe), many situations in which media officials face moral choices. Consider three examples. If NBC produced a documentary on the secret philandering of recent presidents, can anyone doubt that ratings would be fantastically high? Why is NBC unlikely to sponsor such a show? Not because laws prevent it, or because the facts would not be true, but because it would be in terrible taste. In the long run it would probably damage NBC's image.

Now a second thought experiment. A few scientists have recently claimed that blacks are slightly inferior to whites in genetically determined intelligence. NBC could produce a show that would permit the advocates of such opinions to present their case, unhampered by equal time for their opponents. After all, in a free society should not all viewpoints be aired, and would not millions of viewers be interested? Of course the film would start with a disclaimer stating that the opinions you were about to hear were not those of NBC, nor had they been accepted by most scientists.

Why does NBC not present such a show? It cannot be because ratings would be low, and I do not think that fear of damage to NBC's image, though it would no doubt be great, tells the entire story. Incredible as it may seem, I believe there are NBC executives who would consider such a documentary to be not only in bad taste but also morally wrong.

Our third example is, alas, not a thought experiment at all. It is a show that NBC actually sponsored in the fall of 1977, a ninety-minute special called *Exploring the Unknown* and featuring Burt Lancaster as narrator. The film opened with a fleeting voice-over pointing out that although the facts to be presented were believed true they had not been "conclusively proven"—a strange disclaimer, because science never conclusively proves anything. No one I spoke to who saw the show could recall the disclaimer. What they did remember was ninety minutes of dramatic glimpses into what Lancaster proclaimed to be great new frontiers of modern science.

They saw a French magician-turned-psychic bend an aluminum bar with the power of his mind. They saw a Japanese boy stare into a Polaroid camera, and when the film was developed, lo! there on the film was a photograph of the Eiffel Tower. A psychic bounced up and down on his back while a British paraphysicist solemnly concluded that the psychic had probably levitated. Psychic surgeons operated on patients without any instruments except their hands, although plenty of blood and tissue seemed to exude from the magic incisions. Someone moved a laser beam by PK. Another produced paintings under the guidance of long-dead artists. Ingenious inventions were displayed; then viewers were told that discarnate spirits had inspired them. Nothing in the film gave the slightest clue that everything being shown was considered rubbish by the entire scientific community.

In my opinion this atrocious film, and dozens of others like it that have been shown on television in prime time during the past few years, was both in bad taste and morally reprehensible. It was wrong not just because it prompted some seriously ill viewers to fly to the Philippines for a worthless "operation" that could kill them if they neglected legitimate medical help, not just because it pandered to the public's hunger for the paranormal, but because it contributed to the growing inability of citizens to tell good science from bad.

Legislators and government officials, as ignorant of science as television producers and directors, reflect this trend. As a result, public funds are often diverted from worthy projects to worthless ones. In Samuel Goudsmit's book *ALSOS* (the code name for a secret project he headed to determine how far the Nazis had gone in atomic research during World War II) there is a sad scene in which Goudsmit confronts his old friend Werner Heisenberg, one of the few top German physicists who collaborated with Hitler. Germany had just been defeated, and Goudsmit listened patiently while Heisenberg spoke proudly about what he and his few assistants had achieved in the field of atomic explosives. Goudsmit could not then tell him how trivial it all was. It was trivial partly because knuckleheaded Nazi officials had squandered money on puerile projects, partly because great physicists had fled the Fatherland.[2]

The Nazi movement flourished at a time of enormous interest among the German people in the paranormal. Astrology boomed as never before. The public became obsessed with strange diet and health fads and bizarre pseudoscientific cults. A good case can be made for the view that this sprouting of crazy science in Germany made it easier for the populace to buy the crackpot anthropology that supported Hitler's "final solution." We are far behind Nazi Germany in such trends, but the similarities are glaring enough to be alarming.

As always with such manias, causes are multiple: the decline of traditional religious beliefs among the better educated,[3] the resurgence of Protestant fundamentalism, disenchantment with science for creating a technology that is damaging the environment and building horrendous war weapons, the increasingly poor quality of science instruction on all levels of schooling, and many other factors. One factor, often overlooked in trying to explain recent mass enthusiasms, is the role of the media as feedback. This has always been true, but the fantastic power of television and motion pictures to influence opinion has made the feedback a force that now rapidly accelerates any trend. Producers of shows, like publishers of books and magazines, are correct in saying that they respond to public demand rather than initiate it. But just as mild porn stimulates a demand for pornier porn, and mild violence a demand for more violent violence, so does crazy science create a demand for crazier science.

Many decades ago, in *Fads and Fallacies in the Name of Science*, I wrote about some of the outrageous parasciences of the fifties, just before the occult revolution got under way.[4] I have tried to cover part of the current scene in my more recent anthology of articles and book reviews, *Science: Good, Bad and Bogus.*[5] Here I shall limit my remaining remarks to the milder claims of parapsychology.

As I made clear in *Fads and Fallacies*, I believe the claims of responsible parapsychologists deserve to be taken more seriously than any of the other topics listed above as prominent in the current occult explosion. Writers of shabby books about the paranormal usually make no distinction in worth between parapsychology and, say, astrology. Even leading parapsychologists have added to this confusion by contributing to periodicals, like *Fate* and *New Realities*, which they themselves regard with contempt. Clearly there is a rough continuum that runs from unorthodox but respectable science at one end to astrology at the other. Everyone agrees that there are no clear-cut criteria for distinguishing good science from bad. But rather than repeat what I and others have written elsewhere about the hazy guidelines for recognizing cranks,[6] I wish now to underscore a simpler point that I made earlier: the difficulty of drawing precise boundaries around portions of continua does not mean it is useless to distinguish widely separated portions of continua.

We can pin down the point with an extreme case. There are still flat-earth societies in parts of the world, including (where else?) California. If the leader of such a group is sincere, is there anyone not a flat-earther who would hesitate to call him a crank? The usefulness of the word is not impaired by the fact that crank science lies at one end of a spectrum that fades into reputable science at the other. The situation is not unlike that in an old joke about a rich man who asked a woman if she would sleep with him for a million dollars. After she agreed, he offered five dollars. The woman was insulted. "What do you think I am, a whore?" "That's been settled," the man replied. "We are now haggling over the price."

No one can pretend that cranks—or prostitutes—do not exist, but trying to decide where to place a given maverick on the spectrum that runs from good science to crackpottery is haggling over degree of credibility. There are no sharp criteria, and even a consensus among top scientists can reflect a wide range of cultural influences that distort judgments. Johannes Kepler is often cited as a great scientist whose views were mixed with astrology and eccentric speculations in contrast to the sounder views of Galileo. Yet even Galileo could not accept Kepler's correct theory of elliptical orbits or his claim that the moon causes tides. Today there are sophisticated techniques, unknown to Galileo and Kepler, for evaluating scientific theories, but they are still far from free of individual and cultural bias. There is always the possibility that a new point of view that seems outlandish to most scientists will eventually turn out to be right.

I am frequently criticized for having included a chapter on Joseph Banks Rhine in my *Fads and Fallacies*. Many who fault me for this do not recall that on the first page of that chapter, I wrote:

> It should be stated immediately that Rhine is clearly not a pseudoscientist to a degree even remotely comparable to that of most of the men discussed in this book. He is an intensely sincere man, whose work has been undertaken with a care and competence that cannot be dismissed easily, and which deserves a far more serious treatment than this cursory study permits. He is discussed here only because of the great interest that centers around his findings as a challenging new "unorthodoxy" in modern psychology, and also because he is an excellent example of a borderline scientist whose work cannot be called crank, yet who is far on the outskirts of orthodox science.[7]

My opinion of Rhine's sincerity remains unchanged, though as more facts come to light about his early work, the less sure I am of his care and competence. My attitude is much the same toward other top parapsychologists. In my opinion they have not established their claims beyond a low degree of credibility, and at times I find them extraordinarily gullible, but I do not regard most of them as either fools or knaves. Let me do my best to make clear what I think of their work, and why I am not a paranormalist even in the limited sense of believing in psi.

The first point to emphasize is that the claims of parapsychology are scientific claims. They have no necessary connection with any religious or philosophical belief. I do not mean that they are not often connected in some way with such beliefs. For example, it is fashionable now to explain the healings of Jesus and the cures of faith healers as a form of psychic healing, and to find support for levitation in the biblical account of how Jesus ascended into the skies or the Catholic doctrine of the assumption of Mary. I mean only that the claims of parapsychology are set forth as resting on empirical evidence, and that one may hold any metaphysical or religious opinion and accept or reject psi solely on scientific grounds.

I am constantly amused by letters from well-meaning souls who assume that my skepticism about psi is a product of my atheism. It is true that I am one of the founders of the Committee for the Scientific Investigation of Claims of the Paranormal, first sponsored by the American Humanist Association, but I am not a humanist in their sense of the term. It also is true that the narrator of my novel *The Flight of Peter Fromm* is a secular humanist. But anyone who reads that curious novel carefully should realize that my purpose was to contrast the views of my narrator with the persistent theism of my young protagonist, and that my narrator's opinions no more reflect my own than the conservative political views of the narrator of *The Late George Apley* reflect the opinions of John Marquand, or the psychoses of the narrator of *Pale Fire* reflect the personality of Vladimir Nabokov. As I shall confess in later chapters, I am not only a theist but also, in a sense, a Platonic mystic. But what do God or the gods have to do with the question of whether PK can bend a spoon? Most of those I know who believe strongly in psi are atheists. Sigmund Freud and Mark Twain come to mind. Both believed in telepathy and both were atheists. And there have been and are a multitude of theists who are as skeptical as I of the claims of parapsychology. The question of whether psi forces exist is a scientific question. It should be as unaffected by philosophical and religious beliefs as the question of whether gravity waves exist.

Skeptics sometimes say that they don't believe in psychic forces but that the world would be more exciting if such forces did exist. Maybe. In some ways it would be exciting if, say, gravity could be suspended by mental effort, as leaders of Transcendental Meditation insist, but in other ways it would create enormous confusion. Jesus once remarked that no man by taking thought can add one cubit to his stature. It is amusing to read about how Alice grows and shrinks as she nibbles opposite sides of the blue caterpillar's mushroom, but in the real world such paranormal phenomena would not be so amusing.

I also value the privacy of my thoughts. I would not care to live in a world in which others had the telepathic power to know what I was secretly thinking, or the clairvoyant power to see what I was doing. It could be one of God's mercies that we mortals are able to communicate with one another only by willfully uttering words or making other kinds of signs. One of Emily Dickinson's unfinished poems begins with lines I have always liked:

A letter is a joy of Earth.
It is denied the gods.

Rhine once wrote about how ESP, if it ever became a reliable human capacity, would be a great force for peace and the control of crime:

> The consequences for world affairs would be literally colossal. War plans and crafty designs of any kind, anywhere in the world, could be watched and revealed. With such revelation it seems unlikely that war could ever occur again. There would be no advantage of surprise. Every secret weapon and scheming strategy would be subject to exposure. The nations could relax their suspicious fear of each other's secret machinations.
>
> Crime on any scale could hardly exist with its cloak of invisibility thus removed. Graft, exploitation, and suppression could not continue if the dark plots of wicked men were to be laid bare.[8]

It seems not to have occurred to Rhine that such awesome powers could be used just as easily by a police state. They would be tools with a far wider scope for repression and terror than the mere tapping of a phone, opening of a letter, or electronic eavesdropping.

Precognition? The advantages and disadvantages are so obvious that I will elaborate only by citing a Hindu myth and by telling another joke.

The myth: The Hindu god Shiva has a third eye, in the center of his forehead, with which he can see the future. His wife, Parvati, is perpetually and understandably annoyed by this ability because whenever they play a dice game, Shiva cheats by knowing the outcome of future rolls of the dice.

The joke: A devout Christian had become, like Samuel Johnson in his declining years, enormously worried over whether he was destined for heaven or hell. After he had prayed for insight, an angel appeared in the room. "I have," said the angel, "good and bad news. First the good news. You are going to heaven."

The man was overjoyed. "And the bad?"

"Next Tuesday."

PK opens up even more terrifying possibilities. I am not enthusiastic over the possibility that someone who dislikes me might have the power from a distance to cause me harm. The other side of white magic is black magic, whether it be the blackness of demons or the blackness of psychic power. A few years ago I attended a conference on psi at New York University. The lectures, by distinguished parapsychologists, were followed by question-and-answer periods. Several disturbed young ladies arose to say that they had enemies who were using PK to cause them terrible accidents. How could they prevent this? The speakers looked embarrassed, and I was impressed by the evasiveness of their replies.

Parapsychologists, eager to get research funds from the government, are correct in saying that if PK exists it could be used for sabotage. A psychic who can bend a key in someone's fist could surely use the same power to trip an explosive or damage computer circuitry. It is because of such fears—in my opinion groundless—that military forces in both Russia and the United States have funded secret research on psi. Again, could it be another of God's mercies that we do not have the ability to alter substances by taking thought?

Consider also the havoc PK would inject into science. Experiments often depend on the pointer readings of sensitive instruments. If an experimenter can unconsciously influence those instruments, causing them to favor strong hopes or fears, then tens of thousands of experiments become dubious. This injects cloudy elements even into parapsychology. Many eminent parapsychologists are convinced that PK can affect random number generators widely used in psi testing. How can one be sure that an experiment, instead of measuring, say, the clairvoyant power of a subject, is not measuring the PK power of the experimenter, combined with precognition, over the randomizer used to select targets?

Paraphysicist Helmut Schmidt performed a famous experiment with cockroaches that seemed to prove that cockroaches had the paranormal ability to give themselves more electric shocks than chance allowed. Since cockroaches presumably want to avoid shocks, Schmidt suggested that, because he hated cockroaches, maybe it was *his* PK that influenced the randomizer![9] Parapsychologists often say that when skeptics observe an experiment, they inhibit psi. This could work the other way around. How can we be sure that the success of a psi experiment is not the result of PK influences by believers associated with the test rather than the psi ability of the subject?

Since it is easy to imagine how psi forces could be beneficial—psychic healing, for instance—I have tried to balance the case by stressing some evils. My own feeling, as I have said, is neutral. I neither hope nor fear the reality of psi any more than I hope or fear the reality of quarks. The reason I do not believe in psi is that I find the evidence unconvincing.

Most people, brainwashed by a constant barrage of pro-psi rhetoric that they are not in position to evaluate, are astonished to learn that the great majority of psychologists, especially experimental psychologists, do not believe in psi. How can the public know that the number of full-time psi researchers in the country is exceedingly small? In 1978, Charles Tart, a widely respected parapsychologist, estimated the number to be about a dozen. Most of them, he added, are self-taught and poorly funded. Some are tireless propagandists. They churn out popular books and appear on talk shows, where they make excellent impressions as intelligent, open-minded scientists, bravely rowing against the currents of orthodoxy. Hack journalists pounce on their work to produce a constant avalanche of shabby potboilers that often earn fortunes for themselves and their conscienceless publishers.

How can the public know that for more than fifty years skeptical psychologists have been trying their best to replicate classic psi experiments, and with notable lack of success? It is this fact more than any other that has led to parapsychology's perpetual stagnation. Positive evidence keeps flowing from a tiny group of enthusiasts, while negative evidence keeps coming from the much larger group of skeptics. William James, in *Memories and Studies*, expressed perplexity over having followed the literature of psychic research for a quarter-century and finding himself in the same state of doubt as at the beginning. More recently philosopher Antony Flew confessed similar sentiments. In 1953, in his *A New Approach to Psychical Research*, Flew argued that the evidence seemed too much to dismiss, yet there were no repeatable experiments that settled the matter. In 1976 he had this to say:

> It is most depressing to have to say that the general situation more than twenty-two years later still seems to me to be very much the same. An enormous amount of further work has been done. Perhaps more has been done in this latest period than in the whole previous history of the subject. Nevertheless, there is still no reliably repeatable phenomenon, no particular solid-rock positive cases. Yet there still is clearly too much there for us to dismiss the whole business.[10]

How do parapsychologists account for this? I have already mentioned a marvelous excuse that I call their catch-22. If an experimenter is a skeptic, they maintain, even if skeptics are present only as observers, the skepticism inhibits the delicate operation of psi. (Because psi forces are said to be independent of distance and time, I'm surprised that parapsychologists have not yet attributed a replication failure to someone a thousand miles away who had doubted a week before that an experiment would be successful.) Catch-22 places the skeptic in a position unique in the annals of science. There is *no way* a skeptic can disconfirm ESP or PK to the satisfaction of a believer. Hence we face the dreary prospect that the next fifty years of psi research will be exactly like the last.

Catch-22 is only one of many excuses constantly invoked to account for replication failures. The subject had a headache or was emotionally upset, conditions in the laboratory were not sufficiently relaxed, there was a personality clash between subject and experimenter, apparatus for the test made a distracting noise, an experiment was too complicated, the weather was too cold, the weather was too hot, the subject (for some mysterious reason, this occurs frequently) lost his or her previous abilities, and there are many others.

When I look over the vast literature of parapsychology I am overwhelmed by the absence of those strict controls that are essential for confirming extraordinary claims. If someone announces that he can clairvoyantly view a scene ten thousand miles away when given no more than the spot's map coordinates,[11] such a fantastic claim calls for far more stringent controls than are needed to determine whether he can juggle five balls or play "Dixie" by rapping his skull with his knuckles. Until parapsychologists can come up with experiments reliably repeatable by skeptics who are both capable and willing to impose extraordinary controls, their results will continue to have only the mildest influence on "establishment" psychology.

David Hume's essay on miracles (section 10 of his *Enquiry Concerning Human Understanding*) should be required reading for anyone concerned with evaluating the wonders of psychic research. Hume writes mostly about biblical miracles, but if for them you substitute recent psychic miracles, everything in Hume's essay seems to have been written yesterday. "The knavery and folly of men are such common phenomena," he declares, "that I should rather believe the most extraordinary events to arise from their concurrence, than admit of so signal a violation of the laws of nature."

Of the many valuable commentaries on Hume's essay I particularly recommend the observations of Charles Peirce in the sixth volume of his *Collected Papers*, and chapter 7 of Thomas Huxley's *Hume*, from which I have taken this chapter's epigraph. After the statement there quoted, and with which I fully concur, Huxley goes on to spell out what it means to say that extraordinary scientific claims demand extraordinary evidence:

> But when we turn from the question of the possibility of miracles, however they may be defined, in the abstract, to that respecting the grounds upon which we are justified in believing any particular miracle, Hume's arguments have a very different value, for they resolve themselves into a simple statement of the dictates of common sense— which may be expressed in this canon: the more a statement of fact conflicts with previous experience, the more complete must be the evidence which is to justify us in believing it. It is upon this principle that every one carries on the business of common life. If a man tells me he saw a piebald horse in Piccadilly, I believe him without hesitation. The thing itself is likely enough, and there is no imaginable motive for his deceiving me. But if the same person tells me he observed a zebra there, I might hesitate a little about accepting his testimony, unless I were well satisfied, not only as to his previous acquaintance with zebras, but as to his powers and opportunities of observation in the present case. If, however, my informant assured me that he beheld a centaur trotting down that famous thoroughfare, I should emphatically decline to credit his statement; and this even if he were the most saintly of men and ready to suffer martyrdom in support of his belief. In such a case, I could, of course, entertain no doubt of the good faith of the witness; it would be only his competency, which unfortunately has very little to do with good faith, or intensity of conviction, which I should presume to call in question.
>
> Indeed, I hardly know what testimony would satisfy me of the existence of a live centaur. To put an extreme case, suppose the late Johannes Müller, of Berlin, the greatest anatomist and physiologist among my contemporaries, had barely affirmed that he had seen a live centaur, I should certainly have been staggered by the weight of an assertion coming from such an authority. But I could have got no further than a suspension of judgment. For, on the whole, it would have been more probable that even he had fallen into some error of interpretation of the facts which came under his observation, than

that such an animal as a centaur really existed. And nothing short of a careful monograph, by a highly competent investigator, accompanied by figures and measurements of all the most important parts of a centaur, put forth under circumstances which could leave no doubt that falsification or misinterpretation would meet with immediate exposure, could possibly enable a man of science to feel that he acted conscientiously, in expressing his belief in the existence of a centaur on the evidence of testimony.

This hesitation about admitting the existence of such an animal as a centaur, be it observed, does not deserve reproach, as scepticism, but moderate praise, as mere scientific good faith. It need not imply, and it does not, so far as I am concerned, any *à priori* hypothesis that a centaur is an impossible animal; or, that his existence, if he did exist, would violate the laws of nature. Indubitably, the organisation of a centaur presents a variety of practical difficulties to an anatomist and physiologist; and a good many of those generalisations of our present experience, which we are pleased to call laws of nature, would be upset by the appearance of such an animal, so that we should have to frame new laws to cover our extended experience. Every wise man will admit that the possibilities of nature are infinite, and include centaurs; but he will not the less feel it his duty to hold fast, for the present, by the dictum of Lucretius, "Nam certe ex vivo Centauri non fit imago" [For it is certain that no image of a centaur can be made from life], and to cast the entire burthen of proof, that centaurs exist, on the shoulders of those who ask him to believe the statement.

Defenders of parascience often refer to its sensational discoveries as "white crows." It is a poor figure of speech, because white crows are no more unusual than black swans or blue apples. Huxley's choice of a centaur is a much better metaphor. There is not much difference in degree of credibility between a centaur and the wee folk with gauzy wings that Conan Doyle believed had been photographed by two girls in an English glen. The Uri Geller metal-bending effect, the movements of objects by Nina Kulagina in Russia and Felicia Parise in the United States, the claims by Harold Puthoff and Russell Targ that there is an easy way to use precognition for winning at casino roulette,[12] the "thoughtography" of Ted Serios, reports of UFO close encounters of the third kind, and a thousand other marvels that have excited not just the hack writers but many distinguished parapsychologists—these claims are much closer to ancient reports of centaurs than any reports of white crows.

Where in my opinion do parapsychologists go wrong? There is no single answer. In most cases I believe their results are the product of unintentional bias in designing experiments and analyzing raw data. In all types of research that depend heavily on statistics it is easy for experimenters to find what they passionately desire to find. It has often been pointed out that as Rhine learned more and more about controls, his results became less and less impressive. He never found a subject to rival Hubert Pearce, who once called twenty-five ESP cards in a row correctly. He never found another horse like Lady Wonder that could read his mind. Early issues of his journal are filled with papers revealing such trivial controls that if those same papers were submitted now to the same journal they would be rejected. It is no credit to today's parapsychologists that they continue to praise these early, strongly flawed experiments.

Another explanation of many classic psi tests is simply clever—sometimes not even clever—cheating on the part of subjects. As I have often said, electrons and gerbils don't cheat. People do. There is a type of individual, extremely common in the history of psychic research, who has no financial motive for cheating but does have a strong emotional drive to cheat. Such persons get their kicks from being considered psychic by their parents, by their friends, by parapsychologists, and by the public. Frequently they are young students who have the added incentive of wanting to please a teacher and to get good grades. In many cases they firmly believe they have paranormal powers. To reinforce and magnify what they think are authentic talents, they resort to cheating.

The history of spiritualism swarms with mediums who had just such twisted personalities. Robert Browning's long poem "Mr. Sludge the Medium" is about one such man, D. D. Home. Handsome, intelligent, urbane, Home was the greatest medium of his day, perhaps of all time. I do not know what Home's private beliefs were about the dead, but I think Houdini described him accurately when he called him "a hypocrite of the deepest dye." Home even had the gall to write a book on methods by which other mediums cheated, carefully leaving out, of course, his own methods. But most of today's self-styled super-psychics live drab lives and have little talent for public entertainment. The only way they can get themselves noticed is by performing dull magic tricks and pretending they are not tricks. Some become consummate actors who play the role of simpleminded, sincere souls, utterly at a loss to comprehend their peculiar "gifts."

How often did Doyle and other Spiritualists declare that Mrs. So-and-So could not possibly have cheated when her trumpets floated about the darkened room because she was such a sweet, innocent grandmother who obviously knew nothing about magic? Sweet and innocent, my foot! The crafty old lady usually had behind her fifty years of sordid experience as a professional mountebank. No successful card hustler acts like a hustler when he is playing with strangers. No quack doctor talks or behaves like a quack. Psychic charlatans never talk or act like charlatans.

Finally, there are those rare cases when a parapsychologist himself cheats. A sad recent instance is the case of Walter J. Levy, Jr., the director of Rhine's laboratory who had been chosen by Rhine to be his successor. When *Time* devoted a cover article (March 4, 1974) to the occult explosion, its science editor, Leon Jaroff, told me that the most sensible letter of protest came from Levy. A few months later Levy was caught flagrantly cheating by pulling a plug on a counting device to make it register an abnormally high number of hits.[13] I have it on good authority that his crime was detected not because of careful controls in Rhine's laboratory, but because suspicious young staff members set a careful trap for him.

A more recently uncovered case of cheating, by a much more distinguished parapsychologist, is that of Samuel G. Soal. At first Soal was highly scornful of Rhine's work, especially the dice tests. Suddenly he began to get positive ESP results, and quickly became England's most famous psi experimenter. "Soal's work was a milestone in ESP research," said Rhine. "Entirely apart from the intrinsic value of his work, which is high, the manner in which his conclusion forced him to reverse his position on ESP has given it additional status." The evidence that Soal faked the data in one of his most widely respected tests is now overwhelming.[14]

To summarize, in my view there are three major sources of error in the classic psi experiments: unconscious experimenter bias, deliberate fraud on the part of subjects, and occasional fraud on the part of investigators. It is possible, of course, that the world is on the verge of a great new Copernican Revolution of the mind. I cannot say that psi forces do not exist. I do say that the evidence for them is feeble. Extraordinary claims demand much more extraordinary evidence than parapsychologists have been able to muster. When experiments can be reliably replicated by skeptics, when it is evident that controls are commensurate with the wildness of the claims, and when knowledgeable magicians participate in the designing and witnessing of such experiments, I will not hesitate to change my mind.

7. NEW THOUGHT, UNITY, AND ELLA WHEELER WILCOX

Ella Wheeler Wilcox is almost totally unremembered these days except for a few lines of verse often quoted by persons who don't know their origin. In her day she was probably the most popular, most loved poet in America. She was an ardent promoter of a movement called New Thought, which was very similar to the later movement called New Age. New Thought had close ties to Christian Science, which is why I devoted a chapter to Wilcox in my *Healing Revelations of Mary Baker Eddy* (Amherst, NY: Prometheus Books, 1993), my critical biography of the founder of Christian Science.

I consider Wilcox a cut above Christian Scientist Edgar Guest. Although her religious opinions were as nutty as Edgar's, some of her verse in my opinion is still worth reading.

Because I want to focus on the support given to New Thought by the poet Ella Wheeler Wilcox (1850–1919), I will make only a feeble attempt to summarize the extraordinarily complex history of the New Thought movement. It was foreshadowed by the New England transcendentalists, especially by the writings of Emerson, but its more immediate debt was to the teachings of Phineas Parkhurst Quimby and Mary Baker Eddy. Hundreds of famous men and women were caught up in New Thought. Thousands of books were devoted to the movement and tens of thousands of magazine articles.*

The movement began slowly with independent local groups popping up here and there, usually organized by an energetic leader such as Mrs. M. E. Cramer, who formed an early group in San Francisco called the Divine Science Association. In 1890 a New Thought convention in Hartford, Connecticut, led to the founding of the first major national organization, the International New Thought Alliance (INTA).

I would have imagined, before I researched this chapter, that INTA had long since faded, but no, the organization, headquartered in Mesa, Arizona, is still very much alive. It publishes *New Thought*, a handsome quarterly that began in 1914. Blaine C. Mays is its present editor. Its forty-four-page Summer 1992 issue announced the seventy-seventh INTA Congress in July at the Riviera Hotel in Las Vegas. There are ads for scores of New Thought books and tapes, and for a New Thought cruise to the Caribbean. Churches advertising in the issue have such names as the Hillside International Truth Center, Atlanta; the First Church of Religious Science, Manhattan; the First Church of Understanding, Roseville, Michigan; the Church of the Universal Christ, Baltimore; the First Church of Divine Science, Milwaukee; the Unity Church of Christianity, Houston; the Huntington Beach (California) Church of Religious Science; the Sanctuary of Truth, Alhambra, California; and the Unity Church, Hammond, Indiana. Four pages of small type list almost two hundred other New Thought churches and centers around the nation. You can even buy T-shirts with the words ONLY LOVE IS REAL on one side, and ONLY LOVE HEALS on the other.

New Thought was at first confined mainly to a pantheistic emphasis on God as permeating all of nature and dwelling inside every person. By getting "in tune with the infinite" (to quote the title of one of

*The best and most accurate history of New Thought is Charles S. Braden's 571-page *Spirits in Rebellion: The Rise and Development of New Thought* (Southern Methodist University Press, 1963). Its ninth printing (1984) is currently available as a paperback. In the back of his book Braden lists more than a hundred New Thought periodicals that came and went, estimating that his list comprises only about a third of the total number.

the earliest and certainly the most popular of all New Thought books, and written by Ralph Waldo Trine) one could not only free oneself from disease without using drugs, but one would become supremely happy and make lots of money. Although early New Thoughters shared the Christian Scientists' refusal to accept medical treatments, they soon diverged from Eddy by recognizing the physical basis of ailments and becoming much more tolerant of doctors.

Jesus was always considered a prophet of New Thought, but the movement, like Christian Science, was essentially non-Christian in its denial of a Fall of Man, and in its consequent assertion that there is no necessity for a blood atonement by Christ or for a hell of eternal punishment for the unsaved. Over and over again in New Thought literature, as in Christian Science and in today's New Age books, the word "atonement" was spelled "at-one-ment" to suggest that each of us is a part of God, united to everything else in the universe. Recall Shirley MacLaine's chant of "I am God" on the beach in a movie based on one of her many autobiographies?

Slowly during the first few decades of the twentieth century, as Charles Braden puts it, New Thought began to behave like the general who jumped on his horse and rode off in all directions. It became as hard to pin down the movement's precise beliefs as it is hard today to say exactly what New Agers believe. New Thought began to absorb all the occult elements that are part of today's New Age. Leaders began to import doctrines from the East, especially reincarnation. There was nothing about reincarnation in the teachings of Jesus, or in the writings of Quimby, but Charles and Myrtle Fillmore, who founded the New Thought church of Unity, enthusiastically embraced the doctrine. We shall have more to say about Unity later. Today, most members of this still flourishing cult believe in reincarnation.

New Thought also began to beat the drums for all forms of psi phenomena (ESP, PK, precognition), numerology, astrology, and (in a few cases) the wonders of the Great Pyramid. The Psychic Research Company in Chicago was one of many publishing houses specializing in New Thought books. It also founded a magazine called *New Thought* (not the later INTA publication) of which the poet Ella Wilcox was an associate editor. Like today's New Agers, New Thoughters became fascinated by such alternative medicines as osteopathy, homeopathy, and naturopathy.

Here is a typical Psychic Research Company ad for a New Thought book by William Walker Atkinson:

A wonderfully vivid book answering the questions: Can I make my life more happy and successful through mental control? How can I affect my circumstances by my mental effort? Just how shall I go about it to free myself from my depression, failure, timidity, weakness and care? How can I influence those more powerful ones from whom I desire favor? How am I to recognize the causes of my failure and thus avoid them?

Can I make my disposition into one which is active, positive, high strung and masterful? How can I draw vitality of mind and body from an invisible source? How can I directly attract friends and friendship? How can I influence other people by mental suggestion? How can I influence people at a distance by my mind alone? How can I retard old age, preserve health and good looks? How can I cure myself of illness, bad habits, nervousness, etc.? (Ella Wheeler Wilcox, *The Heart of the New Thought*, p. 94)

Another ad has this to say about a book by Sydney Flower:

Although this is the last of this series of books it is in some respects the most important of any. A life-time of study and practice will not exhaust its stores of knowledge. It deals with Psychometry, Phrenology, Palmistry, Astrology, Mediumship and Somnopathy. This last is a new word, coined by the author, Sydney Flower, to define his discovery of a new method of educating the young, i.e., during natural sleep. Of this method, a lady writing in the *Washington Post*, of recent date, said: "I never punish my little ones, I simply wait till they are asleep, and then I talk to them, not loud enough, you understand, to wake them, but in a low voice. I tell them over and over that they must be good, I suggest goodness to them, for I think the mind is just as susceptible to suggestion during the natural sleep as during the working state. I concentrate my mind on it, and I am confident that before long all mothers will adopt my method. It is the best way I know of to bring up children." This method is fully described by its discoverer in this work, and the endorsements of prominent physicians are given in full. (Ella Wheeler Wilcox, *The Heart of the New Thought*, p. 96)

It was the Psychic Research Company that published Wilcox's first book about the movement, *The Heart of the New Thought* (1902). The copy I own is identified as a twentieth printing! She followed it with *New Thought Common Sense and What Life Means to Me* (1908), *The Art of Being Alive: Success Through New Thought* (1914), and *New Thought Pastels* (1907), a book of New Thought poetry.

Here is a typically forgettable poem from *New Thought Pastels*:

All sin is virtue unevolved,
Release the angel from the clod—
Go love thy brother up to God.

Of this great trinity no part deny,
Affirm, affirm the Great Eternal I.

When the great universe was wrought,
To might and majesty from naught,
The all creative force was—
                              Thought

Wilcox's first book about New Thought consists mainly of admonitions and platitudes. They are even duller than those in books by liberal Protestant ministers such as Norman Vincent Peale and Robert Schuller, both of whom are carrying on the New Thought tradition. "Feel-good" ministers is what traditional Christians like to call them. Their messages about "the power of positive thinking" and how such thinking will make you happy and prosperous, although clothed in Christian terminology, are straight out of the New Thought movement.

Here are some of Wilcox's gems from *The Heart of the New Thought*:

Age is all imagination. Ignore years and they will ignore you.

• • •

Eat moderately and bathe freely in water as cold as nature's rainfall. [Hydrotherapy, or water cure, was then popular among New Thoughters.]

• • •

Be alive, from crown to toe.

• • •

Regard any physical ailment as a passing inconvenience, no more.

• • •

Never for an instant believe you are permanently ill or disabled. Think of yourself as on the threshold of unparalleled success. A whole, clear, glorious year lies before you! In a year you can regain health, fortune, restfulness, happiness!

• • •

Push on! Achieve, achieve!

The essence of New Thought, Wilcox tells us, is simply the "science of right thinking." By putting yourself in tune with the infinite God who is also within you, "physical pains will loosen their hold, and conditions of poverty will change to prosperity." This belief that right thinking will bring fiscal rewards runs through all New Thought literature from the beginning on to the inspirational books of Schuller and Peale. Here is "Assertion," a poem in Wilcox's book *Poems of Power* (p. 18)—by "power" she meant the power of right thinking to tune in to God—that captures the heart of New Thought:

I am serenity. Though passions beat
  Like mighty billows on my helpless heart,
I know beyond them, lies the perfect sweet
  Serenity, which patience can impart.

And when wild tempests in my bosom rage,
"Peace, peace," I cry, "it is my heritage."

I am good health. Though fevers rack my brain
*   And rude disorders mutilate my strength,*
A perfect restoration after pain,
*   I know shall be my recompense at length,*
And so through grievous day and sleepless night
"Health, health," I cry, "it is my own by right."

I am success. Though hungry, cold, ill-clad,
*   I wander for awhile, I smile and say,*
"It is but for a time—I shall be glad
*   To-morrow, for good fortune comes my way.*
God is my father, He has wealth untold,
His wealth is mine, health, happiness and gold."

Like all New Thoughters and New Agers, Wilcox is down on traditional Christianity. Its emphasis on sin, hell, and redemption is seen as
blaspheming the true teachings of Jesus about a loving God. Away
with this old and dismal set of doctrines that came from Saint Paul,
not from Jesus! "A wholesome and holy religion," Wilcox writes in *The
Heart of the New Thought*, "has taken its place with the intelligent
progressive minds of the day, a religion which says: 'I am all goodness, love, truth, mercy, health . . . I am a divine soul and only good
can come through me or to me' . . . This is the 'new' religion; yet it is
older than the universe. It is God's own thought put into practical
form" (p. 34).

From *Every-Day Thoughts* (p. 101) I take the following poem:

Let there be many windows to your soul,
That all the glory of the universe
May beautify it. Not the narrow pane
Of one poor creed can catch the radiant rays
That shine from countless sources. Tear away

The blinds of superstition; let the light
Pour through fair windows broad as truth itself
And high as God.
                    Why should the spirit peer
Through some priest-curtained orifice, and grope
Along dim corridors of doubt, when all
The splendor from unfathomed seas of space
Might bathe it with the golden waves of Love?
Sweep up the débris of decaying faiths;
Sweep down the cobwebs of worn-out beliefs,
And throw your soul wide open to the light
Of Reason and of Knowledge. Tune your ear
To all the wordless music of the stars
And to the voice of Nature, and your heart
Shall turn to truth and goodness, as the plant
Turns to the sun. A thousand unseen hands
Reach down to help you to their peace-crowned heights,
And all the forces of the firmament
Shall fortify your strength. Be not afraid
To thrust aside half-truths and grasp the whole.

Like Eddy, Wilcox is convinced that right thinking is the best way for fat people to reduce. She recommends, however, that this be combined with eating only two meals a day, and if one's life is sedentary, one meal is enough. You must be patient. Miracles are seldom instantaneous. They take time.

Near the end of *The Heart of the New Thought* Wilcox admits that not all misfortunes are caused by wrong thinking, or can be done away with by right thinking. She gives as an example the child who "toddles in front of a trolley car and loses a leg," or the person born deaf, blind, or deformed. To account for such evils "we must go farther back, to former lives, to find the first cause of such misfortunes" (p. 79). In her earlier book, *Every-Day Thoughts* (1901), Wilcox was just as specific about reincarnation. "I believe the spirit of man has always existed and always will exist; that it passed through innumerable forms and phases of life, and that which it leaves undone in one incarnation must be accomplished in another."

Deep-breathing exercises are recommended as valuable aids to meditation. One should perform them sitting in a chair facing east in the morning and west at night "because great magnetic forces come from the direction of the sun."

Prenatal influences on the unborn are taken for granted. Did not Napoleon's mother, Wilcox asks, read Roman history while she was bearing him? Wilcox deplores the ignorance of young women about the importance of having happy thoughts during pregnancy; otherwise, fears and worries can damage the baby's brain. Right thinking can even "wear away the stone" of harmful genetic tendencies.

I will spare the reader quotations from Wilcox's later books on New Thought or from the syndicated columns she contributed to William Randolph Hearst's chain of newspapers. But I cannot resist quoting in full an advertisement at the back of *The Heart of the New Thought* (p. 93):

> The many friends and admirers of Ella Wheeler Wilcox will be interested to learn that this gifted author and thinker has connected herself, in the capacity of associate editor, with the *New Thought* magazine,* and that hereafter her writings will appear regularly in that bright publication, of which the aim is to aid its readers in the cultivation of those powers of the mind which bring success in life. Mrs. Wilcox's writings have been the inspiration of many young men

---

*According to Braden, the magazine *New Thought* mentioned in the ad was founded in 1902 by Sydney Flower. Originally it had been *Hypnotic* magazine. The name was changed in 1898 to *Suggestive Therapeutics*. When Wilcox's book was published in 1902, the magazine merged with the *Journal of Magnetism* and took the name *New Thought*. In 1910 it merged with *Health and Success* magazine, and finally expired. It should not be confused with another periodical called *New Thought*, mentioned earlier as the organ of INTA. That journal began publishing as *New Thought Bulletin*, later appearing under various names and finally becoming a quarterly in 1941.

and women. Her hopeful, practical, masterful views of life give the reader new courage in the very reading, and are a wholesome spur to flagging effort. She is in perfect sympathy with the purpose of the *New Thought* magazine. The magazine is having a wonderful success, and the writings of Mrs. Wilcox for it, along the line of the new movement, are among her best. Words of truth, so vital that they live in the memory of every reader and cause him to think—to his own betterment and the lasting improvement of his own work in the world, in whatever line it lies—flow from this talented woman's pen.

The magazine is being sold on all news stands for five cents. It is the brightest, cleanest and best publication in its class, and its editors have hit the keynote of all sound success. The spirit of every bit of print from cover to cover of the magazine is the spirit of progress and upbuilding—of courage, persistence and success. Virile strength and energy, self-confidence, the mastery of self and circumstances are its life and soul, and even the casual reader feels the contagion of its vigor and its optimism.

Free.—The publishers will be pleased to send a handsome portrait of Mrs. Wilcox, with extracts from her recent writings on the New Thought, free. Address, The New Thought, 100, The Colonnades, Vincennes Ave., Chicago.

Ella Wheeler was born in 1850 on a farm in Johnson Center, Wisconsin. She was a precocious, deeply religious child, believing strongly in God, prayer, and the protection of guardian angels. Her writing career began in her teens with selling essays and verse to the *New York Mercury*, and to periodicals published by the houses of Harper and Frank Leslie.

Young Ella shuttled back and forth between the family's small farmhouse and nearby Milwaukee. For a short while she attended the University of Wisconsin, then called Madison University. Her first job was editor of the literary page of a Milwaukee trade magazine that folded after a few months. In addition to her steady output of verse, Ella began selling short stories here and there. Eventually she turned out some forty books, most of them collections of sentimental, moralizing verse. She also wrote a batch of romantic novels, books of essays on general topics, books promoting New Thought, and two autobiographies, *The Story of a Literary Career* (1905) and *The Worlds and I* (1918). Since she contributed regularly to Hearst newspapers and to *Cosmopolitan* magazine, there must be hundreds of her magazine and newspaper articles that never appeared in book form.

Ella's first book of poems, *Drops of Water* (1872), was devoted to attacks on alcohol at a time when the temperance movement was going full blast. It was *Poems of Passion* (1883), however, that catapulted her to fame. The book was roundly condemned by shocked reviewers, which of course made it a bestseller. By today's standards, it contained not a single pornographic stanza, but it did have such lines as "Here is my body; bruise it if you will," and torrid passages (pp. 14–15) such as:

And on nights like this, when my blood runs riot
With the fever of youth and its mad desires,
When my brain in vain bids my heart be quiet,
When my breast seems the center of lava-fires,
Oh, then is the time when most I miss you,
And I swear by the stars and my soul and say
That I will have you, and hold you, and kiss you,
Though the whole world stands in the way.

Some critics have speculated that such passionate lines, written when Ella was in her early thirties, came entirely from her imagination; that actually she was a shy, inexperienced virgin at the time. I don't believe it. Photographs of Ella, in the frontispieces of most of her books of verse, show her to be a beautiful, fine-figured woman, and her autobiography speaks of a "kaleidoscopic panorama of romances" in her youth, including one suitor's "earnest love making."

Ella turned down several proposals of marriage before she met and fell in love with Robert Wilcox, a businessman who shared her enthusiasms for God, prayer, and angels, as well as her forays into New

Thought and occultism. For several years they lived in Meriden, Connecticut, then moved to a Manhattan apartment where they stayed for nineteen years before settling into a Connecticut house on the Sound. Their only child, Robert M. Wilcox, Jr., died twelve hours after birth.

Ella Wilcox has been called the nation's female Eddie Guest. (It is interesting to know that Edgar Guest [1881–1959], America's most popular versifier during the first half of this century, was a devout Christian Scientist.) This too seems to me unfair. Her verse was several cuts above Guest's, even though none of it was great and today it has been almost totally forgotten except for its continued inclusion in anthologies of popular verse. Her best-remembered lines are the opening lines of "Solitude": "Laugh and the world laughs with you; / Weep, and you weep alone." The poem first appeared in the *New York Sun* (February 21, 1883), and later in *Poems of Passion*. The first two lines became so famous that today most people who recall them think they are an anonymous folk proverb. They have even been parodied: "Laugh and the world laughs with you, snore and you sleep alone."

"Solitude" became the center of one of those insane storms of controversy that frequently dog a poem so popular that it is widely reprinted in newspapers and magazines, but without a byline. Colonel John Alexander Joyce (1840–1915) published in 1885 an autobiography titled *A Checkered Life*, which contained a poem identical to "Solitude" except for a transposition of its last two stanzas. Ten years later, in his revised autobiography *Jewels of Memory*, the colonel described how he came to write the poem in 1863 when he was attached to a Kentucky army regiment. Wilcox was understandably furious. She offered five thousand dollars to anyone who could produce a printed copy of the poem prior to its 1883 newspaper appearance. No copy turned up, but the colonel never ceased to claim authorship. "He is only an insect," Wilcox wrote in her second autobiography, "and yet his persistent buzz and sting can produce great discomfort."

Burton Stevenson, who devotes a chapter to all this in his *Famous Single Poems* (1935), reprints a paragraph from the colonel's first version of his autobiography, *A Checkered Life*, in which he admits being confined for several months to the Eastern Kentucky Lunatic Asylum in Lexington because of his "mania" for building a perpetual motion machine.

It is hard to believe, but Colonel Joyce managed to have published biographies of Edgar Allan Poe, Oliver Goldsmith, Robert Burns, and Abraham Lincoln, as well as worthless collections of prose and verse. The controversy he raised over "Solitude" might have died quickly had not Eugene Field (1850–1895), as one of his many practical jokes, kept devoting his newspaper columns to defending Wilcox against Joyce. Wilcox was not amused.

Almost as well known as the opening of "Solitude" are the first four lines of "Worth While," from *Poems of Sentiment*:

> *It is easy enough to be pleasant,*
>   *When life flows by like a song,*
> *But the man worth while is one who will smile,*
>   *When everything goes dead wrong.*

It is true that the above lines sound as if Eddie Guest had penned them, but Guest could never have written "The Winds of Fate," one of Wilcox's most frequently anthologized poems:

> *One ship drives east and another drives west*
>   *With the selfsame winds that blow.*
>   *'Tis the set of the sails*
>   *And not the gales*
> *Which tells us the way to go.*
>
> *Like the winds of the sea are the ways of fate,*
>   *As we voyage along through life:*
>   *'Tis the set of a soul*
>   *That decides its goal,*
> *And not the calm or the strife.*

Nor could Guest have composed sonnets as well crafted as "Winter Rain," from *Maurine and Other Poems* (p. 145):

Falling upon the frozen world last night,
  I heard the slow beat of the Winter rain—
  Poor foolish drops, down-dripping all in vain;
The ice-bound Earth but mocked their puny might,
Far better had the fixedness of white
And uncomplaining snows—which make no sign,
But coldly smile, when pitying moonbeams shine—
Concealed its sorrow from all human sight.
Long, long ago, in blurred and burdened years,
  I learned the uselessness of uttered woe.
  Though sinewy Fate deals her most skillful blow,
I do not waste the gall now of my tears,
But feed my pride upon its bitter, while
I look straight in the world's bold eyes, and smile.

As a good Christian Scientist, Guest would have been incapable of poking fun, as Wilcox did in "Illusion" (from *Poems of Power*, p. 17), at Eddy's belief that everything is *maya*, a dream, except God, the one sole reality:

God and I in space alone
  And nobody else in view.
"And where are the people, O! Lord," I said,
"The earth below, and the sky o'er head
  And the dead whom once I knew?"

"That was a dream," God smiled and said,
  "A dream that seemed to be true.
There were no people, living or dead,
There was no earth, and no sky o'er head
  There was only myself—in you."

"Why do I feel no fear," I asked,
  "Meeting you here this way,
For I have sinned I know full well,

And is there heaven, and is there hell,
   And is this the judgment day?"

"Nay, those were but dreams," the Great God said,
   "Dreams, that have ceased to be.
There are no such things as fear or sin,
There is no you—you never have been—
   There is nothing at all but Me."

Today's critics profess amazement over the once enormous popularity of Wilcox's verse. Why, I don't know. Obviously she was not a great poet. In the prefatory poem of *Maurine* she called herself "only the singer of a little song," awed by the truly great poets of the past, and occupying only a minor place at the edge of poetry's "fair land." She was one of Longfellow's "humbler poets" whose "songs gushed from their hearts." Mediocre as most of her verse is, it is far from doggerel, and in my opinion superior to such nonpoets as William Carlos Williams, so much admired by tin-eared critics who cannot abide musicality in poetry.

Wilcox's resemblances to Shirley MacLaine are striking—one a famous promoter of New Thought, the other a famous promoter of New Age. Both were and are talented, attractive, energetic, unbelievably gullible, and totally ignorant of science. MacLaine's talent is in acting, dancing, and singing. Although Wilcox loved ballroom dancing, swimming, and playing the harp and ukelele, her talent lay in writing poetry and prose.

So great was Wilcox's fame that strange men were perpetually trying to meet and seduce her. "Lunatics I Have Known," a chapter in her 1918 autobiography, is devoted to such creatures, and how she had to seek police protection from some of them. The flames of their passions were fanned by totally false rumors about her many adulterous affairs and previous marriages. The truth is that she and her husband were each married only once, and throughout their lives remained deeply devoted to one another.

Like Mary Baker Eddy and Shirley MacLaine, Wilcox broke away from a conventional Protestant upbringing, but retained an admiration for Jesus as an inspired teacher whose views, she believed, were mangled by his followers. Her early interest in New Thought soon led to her becoming a Theosophist. She speaks in her autobiography of her first encounter with Madame Blavatsky's *Secret Doctrine*, and about her unbounded admiration for Annie Besant, the British theosophical leader. In addition to accepting the Eastern doctrines of reincarnation and karma, she also came to believe in astrology, palmistry, faith healing, astral projection (today called out-of-body experiences), the photographing of thoughts, and all forms of psychic phenomena, including the ability of the dead to contact the living through voice mediums, slate writing, automatic writing, dreams, and Ouija boards.

Wilcox's interest in spiritualism is detailed at length in *The Worlds and I*. She opens this second autobiography with a foreword that consists entirely of an editorial from a spiritualist journal, *Harbinger of Light*, and closes with a paragraph that could have been written by MacLaine:

> Back of all the spheres, at the center of all things, is the Solar Logos—God—from whom all the universe proceeds. In the immensity of space are vast heaven worlds, filled with spirits in various states of development from the earth-bound souls to the great archangels—all bent on returning to the source eventually and becoming "one with God." A wise teacher has said truly, "Orderly gradation is Nature's method of expression. Just as a continuous chain of life runs down from man, so also it must rise above him until it merges into the Supreme Being. Man is merely one link in the evolutionary chain." And Alfred Russel Wallace, who was called the grand old man of science, said, "I think we have got to recognize that between man and God there is an almost infinite multitude of beings, working in the universe at large at tasks as definite and important as any we have to perform. I imagine the universe is peopled with spirits, intelligent beings, with duties and powers vaster than our own. I think there is a spiritual ascent from man upward and onward."

And from this mighty storehouse we may gather wisdom and knowledge and receive light and power, as we pass through this preparatory room of earth, which is only one of the innumerable mansions in our Father's house.

Think on these things.

That last sentence repeats the last four words of Philippians 4:8, a favorable biblical verse among Christian Scientists and New Thoughters:

Finally, brethren, whatsoever things are true, whatsoever things are honest, whatsoever things are just, whatsoever things are pure, whatsoever things are lovely, whatsoever things are of good report; if there be any virtue, and if there be any praise, think on these things.

As an illustration of Wilcox's credulity and ignorance of science, consider the following paragraph from chapter 73 of *Every-Day Thoughts* (p. 249):

Hate is poison. I once visited the laboratory of a scientific man, where, by a peculiar combination of chemicals he was able to test the mental mood of a person who breathed into a glass cylinder. Different mental conditions produced different colors in the chemicals. Anger and resentment produced an ugly brown effect—and in the chemicals thus colored by anger, *a virulent poison* was generated.

This puts a scientific basis to the theory of spiritual-minded people. That hate is poison.

After her husband, Robert, died in 1916, Wilcox drifted into deep depression coupled with a great longing to communicate with him. He had promised that if he predeceased her he would do his utmost to reach her through the veil. Wilcox visited numerous mediums through whom Robert's voice seemed to come, and mediums who produced chalked messages from him on slates. Although Wilcox never doubted the honesty of these seers, or the paranormal phenomena they produced, she believed that evil or malicious spirits in the other world often claimed identities not their own. None of the messages she received impressed her as coming from her husband.

Finally, what she hailed with joy as authentic communications began to arrive—not through mediums, but through her own hands as she placed them, alongside the hands of friends, on a planchette. It scooted rapidly across the Ouija board to spell out messages just as vapid as those she had earlier received through mediums, but so desperate was Wilcox to make contact with her husband that she convinced herself the messages were genuine. Details of this sad story are in Wilcox's second autobiography in two chapters, "The Search of a Soul in Sorrow" and "The Keeping of the Promise."

Most voice mediums in those days spoke only with dead relatives and friends, not with entities who could describe a sitter's previous incarnations. Although Wilcox believed she had had earlier lives, she is silent about them in her books. Had she been living today, she surely would have been as smitten by the New Age channelers as MacLaine and other naive stars of stage and screen. Happily, we are spared details about Wilcox's former lives.

In spite of her scientific illiteracy and other limitations, Wilcox was a fascinating woman, a skillful writer of prose, and a poet of modest talents. Although not active politically or well read in economics, she held democratic socialist views that emerge in many of her essays. Her respect was low for great wealth and conspicuous waste. In *Every-Day Thoughts* she writes: "I believe in co-operative methods of business and in the public ownership of large industries. I have not the kind of brain which formulates the plans for such results, but I have the foresight to see their certain approach" (p. 7).

Wilcox was dismayed by the horrors of war, poverty, and racial injustice, but her positive thinking made her an unbounded optimist. "Slowly but surely the world is gaining a higher moral plane; slowly, but surely, the selfish animal in man is giving way to Man the Image of Divinity . . . The world is growing better with every whirl upon its axis" (*Every-Day Thoughts*, pp. 198–99).

New Thought optimism about the future of humanity was as irre-
pressible as it is in the optimism of Peale and Schuller. Here is another
expression of it in one of Wilcox's perishable poems from *Every-Day
Thoughts* (pp. 192–93):

> *Though the world is full of sinning,*
> *Of sorrow and of woe,*
> *Yet the devil makes an inning*
> *Every time we say it's so.*
> *And the way to set him scowling*
> *And to put him back a pace,*
> *Is to stop this stupid scowling*
> *And to look things in the face.*
>
> *If you glance at history's pages,*
> *In all lands and eras known,*
> *You will find the vanished ages*
> *Far more wicked than our own.*
> *As you scan each word and letter,*
> *You will realize it more*
> *That the world to-day is better*
> *Than it ever was before.*
>
> *And in spite of all the trouble*
> *That abounds on earth to-day,*
> *Just remember it was double*
> *In the ages passed away.*
> *And these wrongs shall all be righted,*
> *Good shall dominate the land,*
> *For the darkness now is lighted*
> *By the torch in Science's hand.*
>
> *Forth from little motes in chaos,*
> *We have come to what we are,*
> *And no evil force can stay us—*
> *We shall mount from star to star.*

We shall break away each fetter
That has bound us heretofore,
And the world to-day is better
Than it ever was before.

Next to Christian Science, the Unity School of Christianity is the largest and best organized of all the religions that trace back to Phineas Quimby and New Thought. Unity operates several hundred "centers" around the nation and abroad, and has a membership of millions. In today's New Age atmosphere it is growing about as fast as Christian Science is declining. Hundreds of books and pamphlets have come from its presses, of which Emilie Cady's *Lessons in Truth* (1894) has been a basic textbook and Unity's most popular work, as well as one of the classics of New Thought. Cady was a homeopathic physician practicing in New York City. Her lessons had originally been published as a series in the magazine *Unity.*

Unlike Christian Science, which it strongly resembles in its pantheism and its emphasis on faith healing, Unity does not accept Eddy's idea that sin, sickness, death, and the material world do not exist. It differs also in embracing the doctrine of reincarnation. Although it professes no dogmatic creed, its husband and wife founders, Mary Caroline (Myrtle) (1845–1931) and Charles Sherlock Fillmore (1854–1948), preached reincarnation, and most of today's Unity members share that belief.

Also unlike Christian Scientists, who have their own churches, Unity members prefer to remain inside Protestant denominations, attending churches of their choice, and at the same time seeking instruction from Unity centers and from its many books and periodicals. The three magazines with the largest circulation are *Unity, Daily Word,* and *Wee Wisdom,* the nation's oldest magazine for children. *Wee Wisdom* even has an edition in Braille.

A graduate of Oberlin College, Myrtle Fillmore began her spiritual pilgrimage as a Christian Scientist, though her contact was not through Eddy's church but through Chicago's Christian Science Theological Seminary. This was a school founded and run by Emma Curtis Hopkins, a former editor of the *Christian Science Journal* who had been excommunicated by Eddy for holding heretical views. After being healed of her tuberculosis by faith, Fillmore founded her own version of Christian Science in 1889. At that time she was living in Kansas City, Missouri, with her husband, Charles. He had been a wealthy real estate developer until he lost a fortune during an economic depression.

After his wife converted him to her brand of Christian Science— perhaps New Thought is a better designation—Charles too was healed, or so he claimed, of a crooked spine, a short leg, and deafness in one ear. He had earlier developed an interest in spiritualism, theosophy, and Hindu mythology, and is said to have been responsible for introducing reincarnation (though not karma, which he rejected) into his wife's thinking. Braden, in his history of New Thought, reveals that Mr. Fillmore once told him that he (Fillmore) was the reincarnation of Saint Paul!

Unity's first magazine was called *Modern Thought*. The title was changed to *Christian Science Thought* in 1890, but when Eddy complained and threatened legal action, the Fillmores shortened the title to *Thought*. Unity was a member of the International New Thought Alliance (INTA) until 1922, when it had grown so large that the Fillmores decided to go it alone.

Charles was so overwhelmed by the dropping of atom bombs on Japan that he wrote an article titled "The Atomic Prayer" that ran in *Unity's* November 1945 issue. The following memorable extract will give you some notion of Fillmore's prose and scientific acumen:

> Our modern scientists say that a single drop of water contains enough latent energy to blow up a ten-story building. This energy, existence of which has been discovered by modern scientists, is the same kind of spiritual energy that was known to Elijah, Elisha, and Jesus, and used by them to perform miracles.
>
> By the power of his thought Elijah penetrated the atoms of hydrogen and oxygen and precipitated an abundance of rain. By the same law he increased the widow's oil and meal. This was not a miracle—

that is, it was not a divine intervention supplanting natural law—but the exploitation of a law not ordinarily understood. Jesus used the same dynamic power of thought to break the bonds of the atoms composing the few loaves and fishes of a little lad's lunch—and five thousand people were fed.

Charles's posthumous book, *The Atom Smashing Power of Mind* (1949), is still available from Unity.

Although the Fillmores rejected Eddy's denial that matter was real, and defended an endless series of reincarnations here on earth, in many respects they agreed with her views. Sin, sickness, and death were considered "unreal" and subject to elimination if one had the right thoughts. Like Eddy, they took the Bible to be a revelation from God, yet a revelation that was not to be taken as historically accurate, but more like an allegory. As in Christian Science, the doctrine of hell was considered blasphemous.

In his later years Charles Fillmore developed a curious aversion toward sex, regarding the act of love as robbing the body of "essential fluids" and hastening old age and bodily decay. Like Eddy, he toyed with the notion that if one could only live up fully to the principles of Unity, the atom-smashing power of the mind might actually smash death and allow one to remain in his or her present body. When asked about this possibility, here is how he replied, as quoted in Ruth Tucker's *Another Gospel* (1989):

> This question is often asked by *Unity* readers. Some of them seem to think that I am either a fanatic or a joker if I take myself seriously in the hope that I shall with Jesus attain eternal life in the body. But the fact is that I am very serious . . .
>
> It seems to me that someone should have initiative enough to make at least an attempt to raise his body to the Jesus Christ consciousness. Because none of the followers of Jesus has attained the victory over this terror of humanity does not prove that it cannot be done.

Like Mrs. Eddy, Charles Fillmore was inspired to write his own version of the Twenty-third Psalm. It could have been written by Norman Vincent Peale or any one of dozens of evangelists who stress the power of God to make one wealthy. Here is the psalm's amazing rewording as given by Fillmore in his book *Prosperity* (1936; p. 60):

The Lord is my banker; my credit is good
He maketh me to lie down in the consciousness of omnipresent
*   abundance;*
He giveth me the key to His strong-box
He restoreth my faith in His riches
He guideth me in the paths of prosperity for His name's sake.
Yea though I walk through the very shadow of debt
I shall fear no evil, for Thou art with me;

Thy silver and gold, they secure me,
Thou preparest a way for me in the presence of the collector;
Thou fillest my wallet with plenty; my measure runneth over.
Surely goodness and plenty will follow me all the days of my life;
And I shall do business in the name of the Lord forever.

The Fillmores made plenty of money, all right, but like Eddy, they lacked enough faith in God's atom-smashing power to live forever in their present bodies. After Myrtle died in 1931, Charles, then seventy-seven, married his longtime secretary Cora Dedrick in 1933. They ruled over Unity with iron fists until both passed on. When Charles died at age ninety-four, his sons took over. More can be read about Charles in Hugh D'Andrade's *Charles Fillmore: Herald of the New Age* (Harper and Row, 1974).

Now in the hands of Fillmore descendants, the cult is headquartered on a vast estate at Unity Village in Lees Summit, Missouri, a suburb of Kansas City. The village includes a large vegetarian cafeteria that is open daily to swarms of visitors, and operates a service called "Silent Unity." The service employs a large staff that is on duty twenty-four

hours each day to provide free consultation by phone (of course dona-
tions are welcome) and to reply to every letter asking for guidance
and help. Oral Roberts was the first Pentecostal evangelist to visit
Unity headquarters and adopt just such a service for his ministry in
Tulsa.

Next in size to Unity is Divine Science, founded a century ago by
Nona L. Brooks and Malinda Cramer. The Divine Light Federation in
Denver sells two books by Brooks: *Mysteries* (1977) and *In the Light of
Healing* (1986). Cramer's *Divine Science and Healing* (1905) seems to
be out of print. Hazel Deane's *Powerful Is the Light: The Story of Nona
Brooks*, published by Denver's Divine Science College in 1945, also
seems to be no longer in print.

The next larger New Thought denomination big enough to support conferences and periodicals is the Church of Religious Science,
founded by Ernest Shurtleff Holmes in 1927. It is still flourishing
with some one hundred branches scattered around the United States.
Books in Print (1991–92) lists more than thirty of Holmes's books and
pamphlets, most of them published by Science of Mind in Los Angeles,
where the church is headquartered. Holmes edited a variety of periodicals, of which the monthly *Science of Mind* is still sold in New Age
bookstores and even on some newsstands. Current issues stress the
paranormal aspects of New Age ideas. Holmes died in Los Angeles in
1960.

Smaller church groups inspired by New Thought have risen and
died over the decades with such names as the Society for the Healing
Christ, Home of Truth, the Church of the Truth, Psychiana (a mail-
order faith), and scores of others. It would take many pages just to cite
the more influential New Thought thinkers of the past and today. Here
are a few listed alphabetically with titles of their best-loved books:

- Raymond Charles Barker, *The Science of Successful Living,
  Spiritual Healing for Today*, and *You are Invisible*;
- Claude Bristol, *The Magic of Believing*;
- Robert Collier, *The Secret of the Ages, Prayer Works*, and *Be
  Rich*;

- Emmet Fox, *Power Through Constructive Thinking, The Sermon on the Mount,* and *The Lord's Prayer;*
- Ervin Seale, *Ten Words That Will Change Your Life* and *Learn to Live;* and
- Ralph Waldo Trine, *In Tune with the Infinite.*

Other poets besides Wilcox have been followers of New Thought, notably Edwin Markham, Victor and Angela Morgan, Don Blanding, and Margery Wilson.

Many Protestant ministers outside fundamentalist and evangelical camps have been strongly influenced by New Thought. We have already mentioned Norman Vincent Peale and Robert Schuller. Incidentally, although Peale posed as a mainline Protestant, he is on record as a firm believer in psychic phenomena of all sorts, including the reality of apparitions of the dead.

In the last few years the most surprising revival of New Thought, in a form closely related to Christian Science and Unity, is *A Course in Miracles,* said to be channeled by Jesus himself through Helen Schucman, a New York psychologist. The *Course's* leading trumpeter is Marianne Williamson, now drawing huge crowds at her lectures in Manhattan and Los Angeles. Her 1992 book *A Return to Love* became such a hot seller that Random House gave her an advance of several million dollars for her next two books.

One of the admirable features of New Thought, at least among most of its leaders past and present, is its tolerance of beliefs other than its own. This is also true of most New Agers. Does not Shirley MacLaine, for example, frequently say that persons are free to do their own thing, to make their own space?

New Thought poet Edwin Markham expressed this tolerance well in "Outwitted," one of his most often quoted poems:

He drew a circle that shut me out—
Heretic, rebel, a thing to flout.
But Love and I had the wit to win:
We drew a circle that took him in!

Ella Wheeler Wilcox shared the same sentiments. Let her have the final word. Here, from *Poems of Power* (p. 159), is "The World's Need":

> *So many gods, so many creeds,*
>> *So many paths that wind and wind,*
>> *While just the art of being kind,*
> *Is all the sad world needs.*

8. WAS THE SINKING OF THE *TITANIC* FORETOLD?

The late Ian Stevenson (he died in 2006), a passionate believer in psi powers and reincarnation, was firmly convinced that Morgan Robertson's sea novel *Futility* was an outstanding case of precognition. Published fourteen years before the *Titanic* disaster, it describes the sinking of a huge ship called the *Titan* that sank in the North Atlantic after it struck an iceberg. I argue here that the parallels in Robertson's novel with the *Titanic*'s sinking are no more than coincidences combined with Robertson's knowledge of the White Star company's plans to build a gigantic ocean liner.

It is highly probable, Aristotle wrote, that improbable events occur. This can be modeled by tables of random numbers. For example, in the unending digits of pi, starting with the 17,387,594,880th digit, there is a sequence of 0123456789. Incredible? Not at all. Somewhere among the infinite digits of pi there is certain to be any specified sequence. It would be more incredible if pi's expansion of digits did *not* contain amazing patterns. Given the enormous number of ways coincidences can occur in one's lifetime, it is surprising how few such events are ever noticed.

The following chapter reprints the introduction to my *Wreck of the* Titanic *Foretold?* (Amherst, NY: Prometheus, 1986), and the preface to its later paperback edition.

INTRODUCTION

There are few persons, even among the calmest thinkers, who have not occasionally been startled into a vague yet thrilling half-credence in the supernatural, by coincidences of so seemingly marvellous a character that, as *mere* coincidences, the intellect has been unable to receive them. Such sentiments—for the half-credences of which I speak have never the full force of *thought*—such sentiments are seldom thoroughly stifled unless by reference to the doctrine of chance, or, as it is technically termed, the Calculus of Probabilities. Now this Calculus is, in its essence, purely mathematical; and thus we have the anomaly of the most rigidly exact in science applied to the shadow and spirituality of the most intangible in speculation.

—Edgar Allan Poe, "The Mystery of Marie Roget"

Poe's lines express an emotion that almost all persons experience at some time in their lives. Over and over again, when the topic of the paranormal comes up in conversation, someone is compelled to recount, often in exasperating detail, an instance from the past when he or she was shaken by a coincidence so extraordinary that it seems impossible to believe it was chance. "How do you explain *that*?" will be flung at any skeptical listener. No amount of talk about probability and statistics is likely to have any effect on the person's convictions. The great coincidence has left an indelible impression. I suspect it would be hard to find a parapsychologist whose interest in the field was not strongly stimulated by one or more such personal experiences.

Before discussing the question central to this strange anthology—Was the *Titanic* disaster foreseen by extrasensory perception?—let us take a look at coincidences in general. The single most important thing to understand is that in most cases of startling coincidences it is impossible to make even a rough estimate of their probability. They are what mathematicians call problems that are not "well formed." Consider, for example, the most common type of precognition—the precognitive dream. There simply is no way to evaluate the degree to which such a dream runs counter to ordinary statistical laws.

Most dreams contain a wealth of vaguely defined, unrelated events. It is impossible to know how many events in a precognitive dream were quickly forgotten because they had no relation to any waking events in the near future. Assume a woman dreams that her Aunt Mary dies in a fire. In the same dream Aunt Mary's husband escapes the fire by jumping out a window and breaks a leg. A few days later one of the following events takes place: Aunt Mary dies of an illness, her husband breaks his arm in an auto accident, or a house in the neighborhood catches on fire. If Aunt Mary dies, the dreamer will be able to tell friends that only a few days ago she dreamed that her aunt died. If the husband breaks an arm, the dreamer may recall that in her dream he broke one of his bones, she isn't sure which, but she *thinks* it was his arm. And of course if a house nearby catches fire she will recall *that* aspect of the dream. Other events in the same dream, of which there could be scores, will be totally unremembered. Even if the dream stressed only the three events mentioned, the mere presence of all three raises the probability of a meaningful correlation.

The probability of at least one meaningful correlation rises even more steeply when we consider the time lag between a precognitive dream and the event. Many such dreams occur several days, some times even weeks, before the dramatic event. This means we have to consider all the events that occurred in all of the person's dreams over what may be a substantial number of days. Because there is no way to retrieve the hundreds of events that may have occurred in dreams during the period preceding the event, the task of estimating the probability of one correlation is hopeless.

There is still more to consider. Every night, all over the world, billions of people are dreaming. Is it not obvious that the probability of remarkable correlations of dream events and future events occurring in *some* of these billions of dreams is extremely high? Of course, whenever an extraordinary correlation does turn up, there is certain to be an intense, unshakable feeling of improbability in the mind of the dreamer.

With respect to dreams about major disasters that make the headlines, we have no inkling of the millions and millions of times that people dream of such a disaster and nothing happens. You can be sure that every time a great liner sets sail or a huge plane takes off with many passengers it is not unlikely that at least one passenger, or a relative of a passenger, will dream of a sinking ship or an airplane crash. Once the ship or plane has arrived safely, who would mention such a dream or even recall it? But if a disaster does occur, such a dream is at once remembered and passed down in families to children and grandchildren. Earthquakes and floods are less likely to figure in disaster dreams for the obvious reason that no one can fix a time frame for such events. One is much more likely to have an unconscious fear that a loved one will die on a specified air flight than a fear that that person will die in an earthquake. Nevertheless, it would be interesting to know, as of course we cannot, how often people in California dream of earthquakes. For all we know there could be thousands every month. Eventually an earthquake is sure to occur. When it does, anyone who dreamed of an earthquake in the near past will remember it and will feel an irresistible impulse to regard the dream as precognitive.

Finally, it is worth remembering that after any major disaster there is a curious type of person, anxious to gain recognition in a community, who will lie about a precognitive dream. If the person is a professional psychic or has a reputation among friends of being psychic, the temptation to fabricate such a dream, or to exaggerate a dream, will be strong. Even among people who are completely honest there will be a tendency to exaggerate without realizing it. After telling about a precognitive dream for the umpteenth time, one no longer recalls the dream's actual details, especially if it occurred many years ago. Dreams are hard enough to remember accurately ten minutes after waking! One is soon recalling not the dream itself but pictures that formed in the mind during previous tellings. The only way a precognitive disaster dream can have evidential value is when its details are written down before a disaster and dated in a way that can be verified, such as being described in a letter or published before the event or stated on a radio or television talk show.

It is extraordinarily difficult for most people to grasp the fact that *some* improbable events are extremely probable, and in some cases absolutely certain. If you buy a ticket for the Irish sweepstakes, the probability that *you* will win is extremely low. But the probability that *someone* will win is certain. If you lose, you have no difficulty understanding why. But, if you win, the impulse to attribute this good fortune to something paranormal is hard to resist. Perhaps God answered a prayer. Perhaps the ticket's number had some special meaning—its digits coincided with a phone number, a birthdate, a house number, a zip code, part of your Social Security number, a number you dreamed about, or a dozen other possibilities. If your number differs in only one digit from the winning number, you may feel that the Fates are going out of their way to tease you. The impression will be that you came extremely close to winning, whereas, in fact, every losing number was just as "close" to winning as yours.

Whenever there are a multitude of possible ways in which a coincidence can occur, the occasional appearance of a strong coincidence is not surprising. This frequently happens in scientific investigations. One of the most difficult of all problems involving scientific method is finding good ways to evaluate unusual patterns of data to determine if they are based on a law of nature or are no more than the normally expected anomalies of random coincidence.

Sometimes what is believed to be a coincidence turns out not to be. An excellent example of this is the way the eastern coasts of the Americas seem to fit the western coasts of Europe and Africa. Geologists thought this a coincidence until evidence became overwhelming that a single continent *had* split and its two sides had drifted apart. When the atomic weights of elements were found to be exact multiples of the atomic weight of hydrogen, some chemists thought this a coincidence. It turned out not to be. It seemed a remarkable coincidence that the gravitational mass of a falling object exactly equals its inertial mass until the equivalence was explained by Einstein's general theory of relativity.

Cases like these, and there are countless others, are balanced by cases in which a startling coincidence is really nothing but coincidental. My favorite example is the fact that the sun's disk as it appears in the sky is almost exactly the same size as the moon's. Moreover, the moon goes around the earth in about thirty days, and the sun rotates on its axis in about the same period. The diameter of the earth's orbit is about 186,000,000 miles, and light travels at close to 186,000 miles per second. Taken in isolation, anomalies like these seem remarkable. Taken in the context of the billions of ways that scientific data can have accidental correlations, they are unremarkable.

A failure to appreciate the frequency of unusual patterns, when the number of possible patterns is large, has an obvious bearing on the claims of parapsychology. It has been pointed out many times that enormous numbers of ESP tests go unreported because the results are negative. In the light of the total number of such tests made around the world, one would expect a certain proportion, by laws of chance alone, to show unusual correlations. If large numbers of tests go unreported, the illusion that ESP is behind the published anomalies is enormously strengthened.

Professional soothsayers, especially the out-and-out charlatans, are well aware of the importance of making many predictions. They know their misses will be quickly forgotten and their hits widely publicized. Studies of the thousands of predictions that have been made by Jeane Dixon show such a poor record of hits that I'm surprised no one has suggested that "negative psi" may be the cause. But, when only her successes are listed, the illusion that she can foresee the future is strong. Psychics who claim to solve crimes use a similar technique. They will talk at police headquarters for hours, rattling off a hundred "impressions" about the crime. If among them there are lucky hits, naive police officers and gullible newspaper reporters will be enormously impressed.

Suppose you get a phone call from a stranger who says he knows the winning horse in a forthcoming race. It turns out that the horse wins. Later you get a second call from the same man, giving the winner in another race. He is right again. The third time he calls, he offers to sell you the name of the next winner. Should you buy it? Not if you know what actually has been going on. For the first race, in which seven horses ran, the man called seventy people, taking their names at random from a phone book. The first ten were given the name of horse A; the second ten, the name of horse B; and so on. Of course ten people will have been told the winner. Ten horses were in the second race. The man then called the ten who got the correct name for the first race. He gives each the name of a different horse. Of course one person got the winning name. Now he calls this person a third time with an offer to sell. Not knowing the background data, you would be inclined to estimate the probability of chance explaining the first two calls as a low 1 in 70.

Science-fiction writers are often given more credit than they deserve for remarkable predictions. H. G. Wells, for instance, as I note in chapter 3, above, in *The World Set Free*, opens with a description of how the atom was split. The chapter is an astonishing prophecy of what actually happened. In addition, Wells has a world war occurring in the mid-forties in which "atomic bombs" are dropped. Considering that novel alone, one might suppose Wells was gifted with paranormal foresight. But Wells made thousands of other predictions in his books, most of which were misses. He saw no future for submarines in warfare, for example, and even in *The World Set Free* he has atom bombs dropped by hand through a hole in the bottom of a plane. Considering the millions of stories about the future that have been written, it is perhaps surprising there are not more lucky hits. Lots of science-fiction writers made the safe prediction that astronauts would one day walk on the moon. So far as I know, only a few writers guessed that the first moon walk would be watched on television back on earth.

With respect to extraordinary coincidences in daily life, there is no doubt that many more occur than are recognized. Unless a coincidence is obvious, we miss it simply by not looking for it. I recall an occasion when my wife and I were driving through a strange town. We mentioned somebody's name a block or two before pulling up to a stoplight. I noticed that the street had the same name as the person we were discussing. How many people bother to notice the name of every side street they pass in a car? If they did, dozens of unlikely word correlations would turn up.

Gilbert Chesterton, in a delightful essay on coincidences (in *Alarms and Discursions*), said it this way:

Life is full of a ceaseless shower of small coincidences; too small to be worth mentioning except for a special purpose, often too trifling even to be noticed, any more than we notice one snowflake falling on another. It is this that lends a frightful plausibility to all false doctrines and evil fads. There are always such props of accidental arguments upon anything. If I said suddenly that historical truth is generally told by red-haired men, I have no doubt that ten minutes' reflection (in which I decline to indulge) would provide me with a handsome list of instances in support of it. I remember a riotous argument about Bacon and Shakespeare in which I offered quite at random to show that Lord Rosebery had written the words of Mr. W. B. Yeats. No sooner had I said the words than a torrent of coincidences rushed upon my mind. I pointed out, for instance, that Mr. Yeats's chief work was "The Secret Rose." This may easily be paraphrased as "The Quiet or Modest Rose"; and so, of course, as the Primrose. A second after I saw the same suggestion in the combination of "rose" and "bury." If I had pursued the matter, who knows but I might have been a raving maniac by this time.

Because of my interest in strange and amusing coincidences (you'll find hundreds in *The Magic Numbers of Dr. Matrix*), I tend to notice them in my own life more than most people do. I can't recall ever having had a hotel or motel room number I could not easily memorize because of some coincidence: it was a prime, or the power of a number, or a sequence of some sort (such as 3, 6, 9), or the first decimal digits of pi or *e* or some other familiar irrational number. I find a note in my files that on November 21, 1979, I opened a water bill that said I owed $21.21. Clearly the odds against this triplet of 21s is high, but there are so many ways meaningless patterns can show up in random numbers that if you start looking for them you'll find them all over the place.

On March 15, 1977, when Jimmy Carter was president, *The New York Times* printed the following story:

James Earl Carter is his name. He wears blue jeans, he's a former Georgia peanut farmer, he attends a Baptist church, and he has a daughter named Amy who goes to a public school. But he does not live in the White House. This Mr. Carter is an electrician whose home now is in the New Orleans suburb of Kenner, La. Yes, "everybody jokes about it, and all I can tell them is that as far as I know I'm not related to *him*," said Mr. Carter.

In one of my Dr. Matrix columns in *Scientific American* I spoofed the notion that a model of the Great Pyramid contained psychic forces. I had my numerologist manufacturing pyramids at a spot near Pyramid Lake in Arizona, and I mentioned that it might do President Nixon some good to go there and sit on one of the lake's pyramidal rock structures. Jeffrey Mishlove, who believes in paranormal synchronicity and almost everything else on the psi scene, mentions this column in his *Roots of Consciousness*, one of the wildest books on the paranormal ever published. Mishlove thinks he has hoisted me on my own petard. "On the very day that Gardner's article reached the public," he writes, "newspapers throughout the country carried pictures of President Nixon visiting the Great Pyramid in Egypt—a suddenly arranged visit presumably unknown to Gardner when he wrote his story! Such a synchronicity seems to embody the message that even the skeptical joker is part of the 'cosmic puzzle.'" Mishlove seems to think I should be embarrassed by this. My own sarcasm, he writes, "is the very model of psychic synchronicity."

Well, we poor skeptics seldom win. I'm surprised Mishlove has never credited me with psychic power since I had Dr. Matrix successfully predict the millionth decimal of pi long before it had been determined by a computer program.

Magicians and fake psychics are skilled in taking advantage of coincidences. There are dozens of ways to duplicate a drawing by trickery, but a psychic occasionally will take a chance and not cheat at all. He will draw a picture of some object—picking something he knows from experience people often select (such as a house, a ship, a cat, and so on). He will put the drawing facedown on a table, then ask a person to draw anything he or she wishes. The psychic's drawing may be correct only about once in twenty times; but, when it is, the person will be convinced for life that the psychic had paranormal powers. Without knowing the background of misses, there is no way to estimate the probability that such a matching is not pure chance.

When I was a sailor in World War II, I sometimes entertained shipmates with card tricks. I often began by removing a card (picking a card often named, such as an ace or a face card) and handing it to someone with the request that he not look at its face until he had called out the name of any card that popped into his mind. If the card was not correctly named I would take it back, glance at it myself (not letting anyone else see the face), exclaim "You're absolutely right!" and then put it back in the deck. This always got a laugh and would be taken as a joke. Of course I was bound to hit more often than one in fifty-two tries. I'll never forget one occasion when a sailor said, "Jack of Hearts," and I asked him to turn the card over slowly. It was the Jack of Hearts. His face turned beet red. No doubt he has since told his grandchildren about the stupendous miracle that a strange sailor showed him more than forty years ago.

In 1977 a Japanese mathematician wrote to me about an astonishing coincidence that had occurred on an educational television program. A distinguished mathematician at Keio University was explaining elementary probability to junior high school students. To illustrate a point he had his assistant toss in the air eight poker chips, each one red on one side and white on the other. All eight fell with the same color up! You can imagine how shaken the professor must have been. The probability of such an event is $\frac{1}{2}^7$ or a bit under .008. If a psychic had been present, he would have taken credit for influencing the chips by psychokinesis, but of course unlikely events like this are certain to happen sometime.

Random events often display what mathematicians call clustering, clumping, or bunching. It explains many examples of synchronicity that Arthur Koestler stresses in his book *Coincidence*. If you fling a thousand beans on a table they will distribute themselves in clumps, and this kind of clumping is often mistaken, both in science and in daily life, as a nonchance pattern. You can demonstrate the effect by shuffling a deck of cards, then spreading the deck to see how the colors are distributed. You'll be surprised at how often there will be a large bunch of adjacent cards, all the same color. An even more dramatic demonstration is to obtain a large supply of tiny spheres of two colors. Mix equal amounts of the two colors together and pour the mixture into a bottle with transparent sides. The pattern you see will exhibit such marked clumping that even physicists may suspect static attraction between spheres of the same color.

Normal clumping can be extremely misleading in statistical research. A town, for example, suddenly shows a high incidence of cancer. Is there something at work in the local area to cause this, or is it just random clustering? Astronomers find a large patch of space where there are no stars, or a long chain of galaxies. Are natural laws at work, or is it just clumping? It is sometimes difficult to tell. At the time I write (fall 1985), there have been an unusual number of airline disasters. All sorts of speculations are going around, but it is probably just normal clumping.

A parapsychologist reports an unusual number of hits (or misses) in a run of a hundred trials. ESP or clumping? It is impossible to tell without knowing how many tests were made, not only by that parapsychologist but also by others around the world. If a thousand scientists, in different parts of the world, toss a hundred pennies in the air, some of the outcomes will show an astonishing number of heads. If only these tests are reported, and we have no knowledge of the others, it will be impossible to make an accurate evaluation of the reported tests. Of course all this applies to statistical research in other areas of science as well. The pitfalls are manifold and subtle. Unless an experiment can be successfully replicated many times, the results may be nothing more than a statistical anomaly. Extraordinary claims for new laws of science demand extraordinary evidence.

With this as background, let us now turn to premonitions of great disasters. Whenever there is a major earthquake, or flood, or fire, or volcanic eruption, or an assassination of a public figure, there are always psychics who will claim to have predicted it, and people who will say they dreamed about it before it happened or had a strong premonition of the event. How much reliance can be placed on such claims?

First we must consider the possibility of fraud. A certain number of psychics will go to any lengths to fake a prediction. When these are discounted, there remain cases where successful predictions actually were recorded. Again, the evaluation of probability is difficult. The problem is not well formed because we don't know how many predictions were recorded in some manner and failed. One by-product of the enormous current wave of interest in the occult is the still-growing number of professional psychics on the scene. They make predictions constantly—on the radio, in letters, on television, in newspapers. It is hardly surprising that, out of the thousands of recorded predictions made every year, some turn out to be remarkable hits. I doubt if a year has gone by in the past few decades that some psychic, somewhere, has not predicted a major California earthquake. When the big quake finally comes, as it must, any psychic who predicted it will get credit, and all his or her misses will be forgotten.

Let us turn to the sinking of the *Titanic*. Although only 1,522 or so people lost their lives, as compared with the tens of thousands in the last great Chinese earthquake, the story of the ship's disaster had many elements that made it unusually newsworthy. The *Titanic* was supposed to be unsinkable. It was a palatial liner on which some of the world's wealthiest people had booked passage for the ship's maiden voyage. A combination of careless acts produced the disaster. The captain ignored warnings of icebergs. The ship's speed was increased to meet its scheduled arrival. The ship was inadequately supplied with lifeboats. The crew was untrained for emergencies. Lookouts were not given binoculars. A radio operator in a nearby ship was asleep and failed to receive the SOS. Another nearby ship did not respond soon enough when its crew learned of the sinking. No person or group could be singled out for blame. More than any other disaster of the time, the *Titanic*'s sinking raised in stark form the old unanswerable question for any theist—why would God allow such a senseless loss of life to happen?

The person who has been the most influential in spreading the view that there were widespread psychic premonitions of the *Titanic* disaster is Dr. Ian Stevenson, professor of psychiatry at the University of Virginia School of Medicine in Charlottesville. Stevenson's answer to the problem of evil is reincarnation. He is best known in occult circles for his many books and articles about efforts to prove, over several decades, that some people have genuine memories of past lives. From his point of view, the law of karma is a law of the universe. Evils that occur are part of the process by which souls evolve upward through a succession of perhaps endless lives. This, however, is not the place to discuss reincarnation. We will consider only Stevenson's two published papers on the *Titanic*.

His first paper, "A Review and Analysis of Paranormal Experiences Connected with the Sinking of the *Titanic*," appeared in the *Journal of the American Society for Psychical Research* (vol. 54, October 1960). He opens by recalling an earlier paper in the same journal (July 1956) titled "Precognition: An Analysis, II," by W. E. Cox. (Cox is best known today for his tireless efforts to prove that Uri Geller's psychic powers are genuine.) In his 1956 paper, Cox reported a survey of twenty-eight serious railway accidents in the United States. The raw data, Cox claimed, show that on the days of the accidents significantly fewer people rode the doomed trains. As Cox reasoned, and as Stevenson reemphasizes, unconscious precognition seemed to be at work causing travelers to defer their trip without being aware of their extrasensory perception of the coming accident.

"A considerable number of apparently extrasensory experiences occurred in connection with the dramatic sinking of the White Star Liner *Titanic* in April, 1912," Stevenson continues. "Some of these were apparently precognitive; others, instances of apparent extrasensory perception contemporaneous with the tragedy." He then proceeds to give a brief history of the disaster, followed by summaries of the "most sub-stantial paranormal experiences connected with the disaster." A foot-note makes clear that he is not convinced that "all the experiences I shall review include paranormal cognition." He thinks some show it, while others do not.

Stevenson's first, and by far his best, example of precognition is the short novel by Morgan Robertson, *Futility*, a small book published in 1898, fourteen years before the *Titanic* sank. Stevenson is impressed by ten ways in which the sinking of Robertson's imaginary ship, the Titan, parallels the sinking of the Titanic. (I shall say no more about the parallels here because I discuss them at length in my preface to The Wreck of the Titan, reprinted below.) Stevenson's Experiences 2 through 8 are of the anecdotal sort. Let us look briefly at each.

Experience 2. A Mr. Middleton cancels his passage on the Titanic after a cable from the United States advises him to do so for business reasons. According to his family and friends, Middleton had two dreams before the disaster in which he saw the liner sink. Because fears of ships hitting icebergs in the North Atlantic were prevalent at the time, dreams of this sort must have been extremely common. In his first dream, Middleton saw the Titanic floating keel upward. As Stevenson mistakenly supposed, the liner sank bow upward. He sees discrepancies like this as characteristic of most precognitive dreams. The event gets distorted in the dream.

Experience 3. A New York woman awakens her husband on the night of the Titanic's sinking to tell him she dreamed her mother was in a crowded lifeboat. The mother had not told her daughter she had booked passage on the Titanic. The mother survived the disaster. This seems impressive until we learn that the account comes from a book called The Mystery of Dreams (1949), in which the author, W. O. Stevens, does not even tell us the names of the mother and daughter! Stevenson is reduced to referring to the daughter as "Mr. Stevens' friend."

Experience 4 is not much better. Mrs. Marshall watches the Titanic sail by her home on the Isle of Wight. She clutches her daughter's hand and says, "That ship is going to sink before she reaches America." This childhood memory of her mother's remark is recalled by her daughter in her book Far Memory (1956).

Experience 5 is Mrs. Potter's dream. There is no vision of water, merely "something like an elevated railroad" with people hanging from it in nightclothes. When Mrs. Potter later saw an artist's rendering of the Titanic going down, she said, "This is just what I saw." This is from her book Beyond the Senses (1939).

Experience 6. The minister of a Methodist church feels compelled to have his congregation sing the hymn "Hear, Father, while we pray to

Thee, for those in peril on the sea." According to *You Do Take It With You*, a book by R. de W. Miller (1955), while the congregation sang this hymn, passengers in the second class dining room of the *Titanic* were singing the same hymn two hours before the ship struck the iceberg.

Experience 7. Mr. Hays, a passenger on the *Titanic*, is reported to have said before the accident that the time had come for a great sea disaster. (This is like saying today that the time has come for an earthquake in California.) Unfortunately, after the iceberg was hit, Mr. Hays's precognitive powers deserted him, because he said, "You cannot sink this boat." A little later he added, "This ship is good for eight hours." The ship sank in less than two hours. Why Stevenson bothered to list this beats me.

Experience 8. A psychic named Turvey is said to have predicted, "a great liner will be lost." He sent his prediction in a letter to a lady who reported it in the spiritualist journal *Light* (June 29, 1912).

Experiences 9 through 12 all involve the famous British journalist and spiritualist W. T. Stead.

These are Stevenson's twelve best cases, and I think anyone who considers them carefully will agree that only the first, Robertson's novel, can be viewed as extraordinary. Stevenson adds a few miscellaneous experiences that are even weaker. A Miss Evans, who drowned in the disaster, reportedly tells someone that "a fortune-teller had once told her to 'beware of the water.'" A crewman deserts the *Titanic* when it stops at Queenstown. "As he did not leave a record of his motives," Stevenson remarks, "we can only surmise that these might have included a foreknowledge, perhaps unconscious, of the forthcoming disaster." A military aide to President Taft, Major Butt, writes a letter on February 12 to tell his sister not to forget where he stored his papers in case the "old ship goes down." Unfortunately, he wrote this before going to Europe on the S.S. *Berlin*, but he returned on the *Titanic* and was among those who perished. The anecdote comes from *The Intimate Letters of Archie Butt*, in two volumes (1930).

Stevenson is aware of how weak most of his twelve cases are. They lack "much that we would wish in the way of further details," especially "contemporary affidavits from witnesses" that recorded premonitions prior to news of the ship's sinking. In fact, Stevenson goes at some length to argue that even the novels by Robertson and Stead can easily be considered cases of what he calls "reasonable inference" rather than precognition. Nevertheless, Stevenson presents his cases as though they add up to significant evidence for ESP. His final conclusion is admirably cautious: "We shall never know, but possibly some of these persons behaved sensibly (as proven by the subsequent sinking) in response to an unconscious precognition while attributing their behavior to an irrational belief."

In spite of these cautions, Stevenson was sufficiently interested in the possibility of psychic perception of the *Titanic* disaster to write a second paper: "Seven More Paranormal Experiences Associated with the Sinking of the *Titanic*" (same journal, vol. 59, July 1965).

Experience 13. In 1919 the *Journal of the American Society for Psychical Research* printed a letter received several weeks after the *Titanic* went down. The writer, not named, claims he had a dream on the night of April 14 in which his deceased father appeared to tell him that a ship had hit an iceberg with much loss of life. "Corroborations are not given," said the author of the article in which the letter was printed, "but a trusted member of the Society knows Mr. M. well and considers him reliable."

Experience 14. Mrs. Henry Sidgwick, an ardent spiritualist (as was her husband, a famous British philosopher), tells in a 1923 article about a letter she received from someone who asked that pseudonyms be substituted for all actual names. The letter (dated July 4, 1912) tells of a friend of the writer who lost a brother on the *Titanic*. It seems that on April 19, a few days after the disaster, the sister of the friend saw in a dream the wife and daughter of the doomed man. They were crying. At the time, it is claimed, the sister who had the dream did not know her brother was on the *Titanic*. Since we don't know the names of anyone involved, and Mrs. Sidgwick is reporting a letter from a woman who in turn is reporting what a friend told her about a dream of the friend's sister, the anecdote seems hardly worth mentioning.

Experience 15. Stevenson prints a letter from an actor in White Cloud, Michigan. The actor recalls a dream by his associate Mr. Black, in which Black said he saw a large ship sinking and hundreds of people drowning. Later that day a telegraph agent tells them about the hundreds who died in the *Titanic* disaster. Stevenson discloses that reports on April 15 were optimistic, and that the telegraph agent could not have known of the drowning of "hundreds." But, Stevenson adds, the agent "may have had some extrasensory perception of the disaster, for which the telegraphic news acted as a kind of nucleus and stimulus for conscious expression." Note how Stevenson, never questioning the reliability of the actor's memory, stretches the anecdote to the limit to account for a serious discrepancy in the story.

Experience 16. When Stevenson was in Brazil in 1962 he met a man who told him about his mother's paranormal perception of the *Titanic*'s sinking. He had heard the story from his father. According to the father, on the night of the disaster his wife had a dream in which she said she saw a big ship called the *Titanic* sink after hitting an iceberg. How did Stevenson corroborate this? Why, he later heard from the informant's sister, who was four years old at the time of the disaster, that the story was accurate!

Experience 17. An Englishwoman writes to Stevenson to tell him about a dream she had when she was fourteen. In the dream she saw a large ship sinking in a waterless park near her home. A few days later came the news that her uncle, an engineer on the *Titanic*, had drowned. Stevenson points out some discrepancies. The woman said they learned of the uncle's death when they saw his picture in the *Daily Mirror* on the morning of April 15. The picture actually appeared on May 4 in an illustrated magazine called *Sphere*. No morning newspaper of April 15 carried the story because the news had not yet reached England.

Experience 18. Stevenson gets a letter from a woman who recalls a premonition she had when she was eleven. Her mother, Mrs. Roberts, was about to sail on the *Titanic* as a stewardess. The daughter begged her mother not to go because she had a strong feeling of disaster. Her mother went anyway, and was a survivor. The woman also told Stevenson that on a later occasion she tried to stop her mother from sailing on a hospital ship. Again the mother did not heed the warning, the ship sank, and she escaped a second time. Stevenson adds that a Mrs.

Roberts was on the *Titanic*'s crew list, but he reports no effort to verify her later escape from the sinking of a hospital ship.

Experience 19. A newspaper obituary of one Colin Macdonald, who died in 1963, said he had refused to join the *Titanic*'s engineering crew because of a premonition of disaster. Stevenson looked up the man's daughter. Yes, she told him, her father had a hunch not to go on the ship.

Weak as these additional cases are, Stevenson seems impressed by the fact that more such cases turn up about the *Titanic* than about disasters involving greater loss of life. Why? He thinks that the "sudden, unexpected anticipation of death generated a greater than usual amount of emotion . . . We have evidence from many other studies of spontaneous cases and from laboratory experiments that strength of emotion is an important feature in . . . extrasensory perception." Stevenson also believes that his nineteen cases show that "sleep for many people provides better conditions for extrasensory perception than does the waking state." Discrepancies are explained by the fact that "in physical perceptions the veridical details become distorted or blended with other details associated in the mind of the percipient."

In spite of skeptical remarks here and there, there is little question that Stevenson thinks his nineteen cases provide impressive evidence for ESP. Here is how he summarizes his opinion in an article on premonitions of disasters in the *Journal of the American Society for Psychical Research* for April 1970. "I was able to assemble a considerable number of corroborated reports suggesting that these nineteen percipients had had extrasensory awareness of the sinking of the *Titanic*. Ten of the cases were precognitive."

I mentioned earlier that Stevenson wondered why there were so many more cases on record of ESP involving the *Titanic* than in connection with worse disasters, including the sinking of the *Lusitania* in 1915. His 1970 paper repeats his earlier expressed opinion that, unlike the *Lusitania* and loss of life in military battles, the *Titanic* sinking was "totally unexpected . . . I suggest that the very unexpectedness of the sinking of the *Titanic* may have generated an emotional shock not present in disasters that are less surprising."

In the same paper Stevenson reports on a study by the English psychiatrist J. C. Barker (*Journal of the Society for Psychical Research,* December 1967) of no fewer than thirty-five precognitive accounts "worthy of confidence" about a disaster in Aberfan, Wales, in 1966, when a slag heap slid down a mountainside, killing 144 people. Stevenson recognizes that anecdotal evidence of this sort is less reliable than laboratory experiments. "Perhaps the best evidence for precognition derives from the experiments of [S. G.] Soal," Stevenson writes. Eight years later it was discovered that Soal had cheated shamelessly on some of his tests, rendering all his research suspect.

My own skepticism about ESP is well known. I shall leave it to readers of this anthology as to whether the evidence for paranormal perception of the *Titanic* disaster is strong enough to support such an extraordinary claim or whether we have here the same familiar blend of unreliable anecdotes with coincidences of the sort that are well within the bounds of normal laws of chance.

As for the amazing mystique that has developed around the sinking of the *Titanic*, it is not hard to understand why it has been so long-lasting. It springs from the ironic juxtaposition of a titanic pride—the belief on the part of everyone concerned that this floating museum of conspicuous waste could not be sunk—and the unexpected suddenness with which that belief was shattered. Unlike so many other disasters, this one could easily have been prevented had there not been such a conflux of human errors. Like the fall of Babylon, the *Titanic*'s sinking can be taken as a symbol of the crumbling of proud empires with their similar mix of the rich, the middle class, and the poor—all going down together. And now for the first time in history it can be taken as a symbol of the sudden fate of the entire human race if a combination of human follies should set off a nuclear war.

I wish to thank an old friend, Russell Barnhart, for assistance in researching several aspects of this book.

Since this book first appeared in 1986, public interest in the *Titanic* has been increasing. New books have appeared with such titles as *The Night Lives On*, by Walter Lord; *Her Name: Titanic*, by Charles Pellegrino; *Titanic: An Illustrated History*, by Don Lynch; *The Riddle of the Titanic*, by Robin Gardiner and Dan van der Vat; *Down with the Old Boat*, by Steven Biel; and many, many others.

More than a hundred books have been written about the ill-fated ship. A hundred songs about the disaster were published within a few years after the ship sank. A CBS miniseries featuring George C. Scott as the ship's captain aired on television in 1996. *Titanic*, a musical, opened on Broadway in April 1997. James Cameron's movie *Titanic*, said to be the eighteenth film about the ship, was released in December 1997. A popular early film on the sinking starring Barbara Stanwyck and Clifton Webb has been revived on television.

As everyone knows, the *Titanic*'s wreckage was found in 1985, and there are even plans to lift it to the surface. Thousands of artifacts have been salvaged, not only gold and silver objects, china, jackets, and jewelry, but also stock certificates, playing cards, and even love letters.

The New York Times reported (April 8, 1997) that, contrary to what has been widely believed, the ship did not sink because of an enormous gash in its hull. It sank because of six narrow slits across supposedly watertight holds. There is no longer any doubt, again contrary to survivor reports, that the liner broke in half as it plunged bow foremost into the icy sea.

Several minor errors in this book's hardcover edition are here corrected. Other mistakes were impossible to remedy without repagination, so let me mention them here. Richard Branham wrote to point out that the photo of the *Titanic* off the coast of France was obviously retouched because it shows an aft funnel (smokestack) belching smoke. This funnel was a dummy, constructed only for show.

Charles P. Chaffe convinced me that "Mayn Clew Garnett," the byline of a story reprinted in this book, was an obvious pseudonym. A "clew-garnet" is a tackle attached to the clew of a square sail on old sailing ships, used to haul the sail up for furling.

July Rehm-Norbo called my attention to a curious coincidence on page 27 that I failed to notice. Morgan Robertson's novella was first published by a firm called Mansfield. The man who reprinted the short novel in 1985 lived in Mansfield, Ohio.

Robertson was a believer in psychic powers. The American poet and spiritualist Ella Wheeler Wilcox, in her second autobiography, *The Worlds and I* (pages 219–21), writes of how struck she was by Robertson's precognition. When the *Titanic* sank, she and her husband were on the *Olympic* headed for England. In England she learned of Robertson's prophetic novel. Curious to know more about the matter, she wrote to the author and received the following reply:

> As to the motif of my story, I merely tried to write a good story with no idea of being a prophet. But, as in other stories of mine, and in the work of other and better writers, coming discoveries and events have been anticipated. I do not doubt that it is because all creative workers get into a hypnoid, telepathic and percipient condition, in which, while apparently awake, they are half asleep, and tap, not only the better informed minds of others but the subliminal realm of unknown facts. Some, as you know, believe that in this realm there is no such thing as Time, and the fact that a long dream can occur in an instant of time gives color to it, and partly explains prophecy.

How, Wilcox wanted to know, could "Mr. Robertson fix on almost the very name which was afterward given to the ill-fated sea monster?" New evidence has recently emerged which makes this coincidence far less astonishing.

Everett Bleiler, in his marvelous reference work *Science Fiction: The Early Years* (Kent State University Press, 1990), reports his discovery of an obscure novel by William Young Winthrop. Nothing is known about Winthrop except that he was born in 1852 and lived in Woodmont, Connecticut. Titled *A 20th Century Cinderella or $20,000 Reward*, Winthrop's novel was published in 1902 by the Abbey Press, a

New York vanity house. Bleiler summarizes the plot, calling the book "a very long amateurish novel" and a "silly bore." I mention the novel here only because it refers to a gigantic ocean liner called the *Titanic* that had been built by England's White Star Company some time before 1920, the year in which Winthrop's science fiction story takes place.

Now the actual *Titanic* was also a White Star liner, although not built until a decade after Winthrop's novel was published! This could be another coincidence (believers in precognition will no doubt consider it another paranormal prophecy even though the fictional *Titanic* does not sink), but does it not suggest the following? It seems to me entirely possible that the White Star Company, as early as 1898, when Robertson wrote his novel, had announced plans to construct the world's largest ocean liner and to call it the *Titanic*.

Even more relevant is a quotation from *Titanic: Destination Disaster* (1987), by John P. Eaton and Charles Hass, which was sent to me by Brenda Bright. The authors reproduce in its entirety the following news item from the September 17, 1892, issue of *The New York Times*. This was six years before Robertson's novel was published:

> London, Sept. 16—The White Star Company has commissioned the great Belfast shipbuilders Harland and Wolff to build an Atlantic steamer that will beat the record in size and speed.
>
> She has already been named *Gigantic*, and will be 700 feet long, 65 feet 7½ inches beam and 45,000 horsepower. It is calculated that she will steam 22 knots an hour, with a maximum speed of 27 knots. She will have three screws, two fitted like *Majestic*'s, and the third in the centre. She is to be ready for sea in March, 1894.

The figures given for the planned liner are very close to those Robertson used for his imaginary *Titan*. The *Gigantic* was to be 700 feet long, with 45,000 horsepower, a speed of 22 to 27 knots, and three propellers. The *Titan* was 800 feet long, with 40,000 horsepower (changed to 75,000 in the book's second printing), a speed of 25 knots when it struck the iceberg, and three propellers.

The *Gigantic* was never built. At the time Robertson wrote his novel the White Star had built the *Oceanic* (1871), the *Britannic* (1874), the *Teutonic* (1889), and the *Majestic* (1889). The company always added -*ic* to the names of its liners. After Robertson's novel was published it would build a second *Oceanic*, a *Celtic*, *Cedric*, *Baltic*, *Adriatic*, *Olympic*, and *Titanic*.

It seems clear now what happened. Knowing of plans for the *Gigantic*, Robertson modeled his ship on this proposed mammoth liner. After the use of such names as *Oceanic*, *Teutonic*, *Majestic*, and *Gigantic*, what appropriate name is left for a giant liner except *Titanic*? Not wishing to identify his doomed *Titan* with the White Star line, Robertson dropped -*ic* from the name. The White Star's later choice of *Titanic* for its 1910 ship was almost inevitable. The company was surely aware of Robertson's *Titan*, but perhaps did not mind adopting a similar name because it was firmly persuaded that its *Titanic* was absolutely unsinkable.

A raft of old poems about the *Titanic* has come to light, enough to make a book-length anthology. One of the worst is four stanzas of doggerel by Chicago writer Ben Hecht, which Walter Lord includes in *The Night Lives On*, a sequel to his memorable *A Night to Remember*. I will inflict on the reader only the first stanza:

> *The Captain stood where a captain should*
> *For the law of the sea is grim.*
> *The owner romped ere his ship was swamped*
> *And no law bothered him.*

A much better poem, though with a regrettable racist line in the second stanza, was written by, of all people, Sir Arthur Conan Doyle. Doyle and George Bernard Shaw clashed over how one should react to the *Titanic* disaster. Their argumentative letters to a British newspaper were reprinted in *ACD*, the journal of the Arthur Conan Doyle Society (vol. 5, 1994, pp. 127–49), with commentary and annotations by Doug Elliot. (I am grateful to Dana Richards for a copy.) Learning that the band played ragtime to keep up the passengers' spirits, Doyle was moved to write the following poem, which he titled "Ragtime!"

Ragtime! Ragtime! Keep it going still!
Let them hear the ragtime! Play it with a will!
Women in the lifeboats, men upon the wreck,
Take heart to hear the ragtime lilting down the deck.

Ragtime! Ragtime! Yet another tune!
Now the 'Darkey Dandy,' now 'The Yellow Coon!'
Brace against the bulwarks if the stand's askew,
Find your footing as you can, but keep the music true!

There's glowing hell beneath us where the shattered boilers roar,
The ship is listing and awash, the boats will hold no more!
There's nothing more that you can do, and nothing you can mend,
Only keep the ragtime playing to the end.

Don't forget the time, boys! Eyes upon the score!
Never heed the wavelets sobbing down the floor!
Play it as you played it when with eager feet
A hundred pairs of dancers were stamping to the beat.

Stamping to the ragtime down the lamp-lit deck,
With shine of glossy linen and with gleam of snowy neck,
They've other thoughts to think to-night, and other things to do,
But the tinkle of the ragtime may help to see them through.

Shut off, shut off the ragtime! The lights are falling low!
The deck is buckling under us! She's sinking by the bow!
One hymn of hope from dying hands on dying ears to fall—
Gently the music fades away—and so, God rest us all!

Lord devotes an entire chapter, "The Sound of Music," to the *Titanic*'s band. "The last moments of the *Titanic*," he writes, "are full of mysteries—none more intriguing than those surrounding the ship's band. We know they played, but little else. Where they played, how

long they played, and what they played remain matters for speculation." As I said in my first introduction, not one of the eight musicians
survived.

The White Star Company, I earlier noted, ended the names of its
liners with -*ic*. Its rival Cunard Company ended its ships' names with
-*ia*. If it had built the *Titanic*, Richard Branham speculated in a letter,
the ship might have been called *Titania*. Branham also pointed out
numerous coincidences involving Robertson's *Titan* and the *Lusitania*.
The *Titan* sank in April of an unspecified year. The *Lusitania* was tor-
pedoed and sunk in May 1915. The *Titan's* captain was named Bryce.
The chief engineer of the *Lusitania* was Archie Bryce. The *Lusitania's*
dimensions were actually closer to those of the *Titan* than were the
Titanic's. Branham, tongue in cheek, makes out as good a case for
Robertson precognizing the *Lusitania* disaster as for anticipating the
Titanic's sinking.

One of the most moving accounts of the *Titanic* tragedy was written by Elbert Hubbard. You'll find excerpts in *Elbert Hubbard of East
Aurora*, a 1926 biography by Felix Shay (pp. 513–20). Hubbard concluded his essay by writing, "One thing sure, there are just two respectable ways to die. One is of old age, and the other is by accident.
All disease is indecent. Suicide is atrocious. But to pass out, as did Mr.
and Mrs. Isador Straus, is glorious. Few have such a privilege. Happy
lovers, both. In life they were never separated, and in death they are
not divided."

On May 7, 1915, Elbert Hubbard and his wife Alice perished in the
sinking of the *Lusitania*.

PART III

MATHEMATICS

9. DRACULA MAKES A MARTINI

This chapter opens with a classic brain teaser, then shows how it can be the basis for two mystifying card tricks. The literature on what magicians call mathematical card tricks, or self-working card tricks, is now vast. If you are interested in learning more about mathematical magic, a good introduction is my book *Mathematics, Magic, and Mystery*, a Dover paperback. The chapter here first appeared in *Isaac Asimov's Science Fiction Magazine* (September 1979).

It's cocktail time, my love," said Count Dracula to his wife. "Shall it be the usual?"

"The usual," said Mrs. Dracula.

The count took from his liquor cabinet a bottle containing one quart of vodka and a smaller bottle containing one pint of human blood. He poured a small quantity of blood into the vodka, shook the bottle vigorously, then poured exactly the same amount back into the bottle of blood. Hence at the finish there was again a quart of liquid in the large bottle and a pint in the small bottle.

Mrs. Dracula was sitting with her back to her husband, but she was watching him in a mirror on the living room wall. The count was following the standard Transylvanian procedure for making a vampire martini.

Assume that when vodka and human blood are mixed, neither alters in volume. After the two operations just described, is there more vodka in the pint of blood than there is blood in the quart of vodka, or less, or are the two amounts the same?

You may have come across this puzzle before in the form of identical glasses, one filled with water, the other with wine. In this variant, however, the contents of the two containers are not alike, nor are we told the amount of liquid that is transferred back and forth.

ANSWER

If you tried to crack this puzzle by algebra, using exact quantities, you probably got into a muddle. There is, however, a ridiculously simple proof that the amount of blood in the vodka must exactly equal the amount of vodka in the blood.

We are told that at the finish there was, as before, one quart of liquid in the large bottle, one pint in the small bottle. Consider the large bottle. It is missing an x amount of vodka. Since it remains a quart, this missing amount must have been replaced by an x amount of blood! Of course the same reasoning applies to the small bottle. If it is missing an x amount of blood, and remains a pint, the missing blood must be replaced by an x amount of vodka. In fact, it doesn't matter in the least how many times varying amounts of liquid are transferred back and forth so long as at the finish there is a quart in one bottle and a pint in the other. Even the bottle sizes are irrelevant. The volume of vodka in the blood must equal the blood in the vodka!

Can you invent a simple card trick based on the same curious principle?

Place the twenty-six black cards of a deck in one pile. Next to it place, say, thirteen red cards. Turn your back and ask someone to take as many cards as he likes from the black pile and shuffle them into the red pile. Then take the same number of cards from the formerly all-red pile and shuffle them into the black pile.

You turn around, massage your temples, and announce that your clairvoyant powers tell you that the number of red cards among the black is exactly the same as the number of black cards among the red.

This must always be the case, and for the same reason given in the solution to the martini problem. If you like, you can let a spectator shuffle the two piles together, then deal twenty-six cards into one pile and thirteen into a second pile. The final result will be the same as before.

Now go back and reread the first part of this feature. What whopping error was made in describing Count Dracula's mixing of the cocktails?

The description said that Mrs. Dracula watched her husband in a mirror. As every reader knows, or should know, vampires don't *have* mirror reflections.

POSTSCRIPT

Hundreds of mathematical card tricks exploit essentially the same principle as the one just described. Here is a good example to try on friends.

Before showing the trick, cut a deck of fifty-two cards exactly in half. Turn over one half, then shuffle the twenty-six faceup cards into the twenty-six facedown ones. When you start the trick, show that the deck is a mixture of faceup and facedown cards, but don't say how many are reversed. Let someone shuffle, then hand you the deck under a table. A moment later you bring out the cards, a half-pack in one hand, the other half in the other, and announce that each half contains exactly the same number of faceup cards! This proves to be correct.

Secret: Under the table, count off twenty-six cards. Turn over either of the half-packs before you put the two pack of cards on the table. Do you see why it works? Before reversing one half, the number of faceup cards in either half must equal the number of facedown cards in the other. Reversing either half turns its faceup cards facedown and vice versa. This makes the number of faceup (or facedown) cards the same in each half-deck.

10. THE FIBONACCI SEQUENCE

The Fibonacci sequence begins 1, 1, 2, 3, 5, 8, 13, 21 . . . Each number beyond the first two is the sum of the preceding two numbers. A generalized Fibonacci sequence is one that starts with *any* two integers.

Fibonacci numbers are the subject of a vast literature. There is even a periodical called *The Fibonacci Quarterly*. A good introduction to the topic is *The (Fabulous) Fibonacci Numbers*, by Alfred Posantier and Ingmar Lehmann (Amherst, NY: Prometheus, 2007).

My article on some little-known aspects of Fibonacci numbers appeared in *The Journal of Recreational Mathematics* (34: 183–90, 2005–06).

I think Fibonacci is fun;
We start with a 1 and a 1.
Then 2, 3, 5, 8,
But don't stop there, mate!
The fun has just barely begun.
　　　　　—Arthur Benjamin

It has been almost two decades since I last interviewed Dr. Matrix at a math conference in Lisbon. (The interview is the last chapter of a collection of my *Scientific American* columns, *Penrose Tiles to Trap-*

door Ciphers.[1] Since then I had completely lost track of the old scoundrel and his half-Japanese daughter Iva. So it was with great surprise and pleasure that I encountered him in October 2006 at a conference on number theory at Stanford University. He was scheduled to lecture on "Little Known Curiosities About Fibonacci Sequences."

Iva was no longer traveling with him. She had married a Japanese magician in 1991, the doctor told me, and was living in Tokyo with her husband and two teenage sons, Irving and Joshua.

Dr. Matrix had noticeably aged. His hair was snow-white, but his emerald eyes were as alert and penetrating as ever, and his gait steady. He handed me a typescript of his lecture. From it and from later conversations, I learned the following bizarre facts.

Consider any four consecutive terms A, B, C, D in a generalized Fibonacci sequence. Then A times D is always one leg of a Pythagorean triple, and twice B times C gives the other leg! For example, consider the first four terms of the simplest Fibonacci sequence, 1, 1, 2, 3. Substituting these values in the formulas produces the familiar 3, 4, 5 right triangle. The procedure, of course, generates an infinity of Pythagorean triples, though not, unfortunately, *all* such triples.

The equation for any four consecutive terms is

$$(A \times D)^2 + (2 \times B \times C)^2 = (B^2 + C^2)^2$$

The equation is easily proved. Dr. Matrix cited no reference for this curiosity, but I checked with an editor of *The Fibonacci Quarterly* and learned it had been published by A. Horadam in *American Mathematical Monthly.*[2]

Dr. Matrix showed on a screen the old changing area paradox illustrated in figure 1. The square has an area of 64 square units. When the four pieces are rearranged to make a rectangle, the area jumps to 65 units! If the pieces are again rearranged as shown in figure 2, the area shrinks to 63!

Note that the line segments of this classic paradox are 3, 5, 8, and 13, four terms of the Fibonacci sequence. A well-known property of this sequence is that if any number n is squared, it equals the product of the two numbers on each side of n, plus or minus 1.

In this case the square has a side of 8 and an area of 64. In the Fibonacci series, 8 is between 5 and 13. Since 5 and 13 automatically

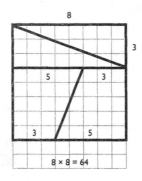

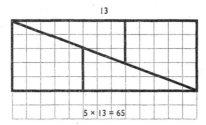

Figure 1.

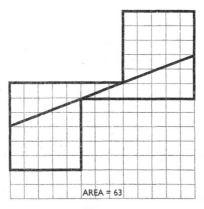

Figure 2.

become the sides of the rectangle, the rectangle must have an area of 65, a gain of one unit.

Owing to this property of the series, we can construct the square with a side represented by any number in the series above 1, then cut it according to the two preceding numbers in the series. If, for example, we choose a thirteen-sided square, we then divide three of its sides into segments of 5 and 8, and rule the cutting lines as shown in figure 3. This square has an area of 169. The rectangle formed from the same pieces will have sides of 21 and 8, or an area of 168. Due to an overlapping along the diagonal of the rectangle, a unit square is lost rather than gained.

A loss of one square unit occurs if we choose 5 for the side of the square. This leads us to a most curious rule. Alternate numbers in the Fibonacci series, if used for the square's side, produce a *space* along the diagonal of the rectangle and an apparent *gain* of one square unit. The other alternate numbers, used for the side of the square, cause an *overlapping* along the rectangle and a *loss* of one square unit. The higher we go in the series, the less noticeable becomes the space or overlapping. Correspondingly, the lower we go, the more noticeable. We can even construct the paradox from a square having a side of only

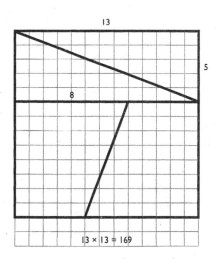

Figure 3.

two units—but in this case the 3-by-1 rectangle requires such an obvious overlapping that the effect of the paradox is completely lost.

The earliest attempt to generalize the square–rectangle paradox by relating it to this Fibonacci series apparently was made by V. Schlegel in *Zeitschrift für Mathematik und Physik*.[3] E. B. Escott published a similar analysis in *Open Court*,[4] and described a slightly different manner of cutting the square. Lewis Carroll was interested in the paradox and left some incomplete notes giving formulas for finding other dimensions for the pieces involved.[5]

An infinite number of other variations result if we base the paradox on other Fibonacci series. For example, a square based on the series 2, 4, 6, 10, 16, 26, etc., will give losses and gains of 4 square units. We can determine the loss and gain easily by finding the difference between the square of any number in the series and the product of its two adjacent numbers. The series 3, 4, 7, 11, 18, etc., produces gains and losses of 5 square units. T. de Moulidars, in his *Grande Encyclopédie des Jeux*,[6] pictures a square based on the series 1, 4, 5, 9, 14, etc. The square has a side of 9 and when formed into a rectangle loses 11 square units. The series 2, 5, 7, 12, 19, etc., also produces losses and gains of 11. In both these cases, however, the overlapping (or space) along the diagonal of the rectangle is large enough to be noticeable.

If we call any three consecutive numbers in a Fibonacci series A, B, and C, and let X stand for the loss or gain of area, then the following two formulas obtain:

$$A + B = C$$
$$B^2 = AC \pm X.$$

We can substitute for X whatever loss or gain we desire, and for B whatever length we wish for the side of a square. It is then possible to form quadratic equations which will give us the other two numbers in the Fibonacci series, though of course they may not be rational numbers. It is impossible, for example, to produce losses or gains of either two or three square units by dividing a square into pieces of rational lengths. Irrational divisions will, of course, achieve these results. Thus, the Fibonacci series $\sqrt{2}$, $2\sqrt{2}$, $3\sqrt{2}$, $5\sqrt{2}$ will give a loss or gain of two, and the series $\sqrt{3}$, $2\sqrt{3}$, $3\sqrt{3}$, $5\sqrt{3}$ will give a loss or gain of three.

Dr. Matrix was generous enough in his lecture to cite as a reference chapters 8 and 9 of my Dover paperback *Mathematics, Magic, and Mystery*[7]—chapters devoted to a variety of surprising geometrical vanishings, including the disappearance of faces and persons! Included is a brilliant breakthrough by amateur magician Paul Curry in which a rearrangement of pieces forms a figure of seemingly the same area but with a large interior hole!

Dr. Matrix closed his lecture by introducing tribonacci numbers. The tribonacci sequence is obtained by summing the previous *three* numbers: 1, 1, 2, 4, 7, 13, 24, 44, 81, . . . In the generalized Fibonacci sequence, the ratio of adjacent terms A and B, when A is divided by B, converges on .618 . . . , the reciprocal of the famous golden ratio. In the tribonacci sequence, the ratio converges on .543 . . . Tetronacci numbers are formed by adding the *four* previous numbers. One can, of course, generalize to *n* terms. As *n* approaches infinity, the ratio of adjacent terms converges on a limit of .518790064 . . .

Such sequences, I later learned from Donald Knuth, Stanford's celebrated computer scientist, were first introduced by Narayana Pandita in 1356, in chapter 13 of a marvelous Sanskrit work titled *Ganita Kaumudi* (Lotus Delight of Calculation).[8] Knuth discusses this and gives further references in volume 4 of his classic *Art of Computer Programming*.[9] The sequence was later rediscovered by Mark Feinberg when he was fourteen. He wrote about it in *The Fibonacci Quarterly*.[10] In 1967, Mark was killed in a motorcycle accident when he was a sophomore at the University of Pennsylvania.

Although it is unrelated to Fibonacci numbers, Dr. Matrix passed along the following improbable curiosity when we were having lunch with Donald Knuth. Arrange the ten digits in *alphabetical* order to form the random, seemingly dull number 8549176320. Divide by 5. The quotient, 1709835264, is another number with each of the ten digits! Divide again by 5. The result is 341967052.8, a third number with all ten digits!

Now divide by 4. You are back to the original alphabetical number, with a decimal point. Do you see why? Dividing twice by 5, then by 4, is the same as dividing by 100.*

I sent this curiosity, discovered by Dr. Matrix, to friend Owen O'Shea, of Cobh (pronounced Cove), Ireland. He is the author of a recently published book, *The Magic Numbers of the Professor*.[11] Owen came back with many surprising properties of the "uninteresting" alphabetical number. For example, its prime factors are 2^{10}, 3^3, 5, and 61843. It follows that 8549176320 is evenly divided by all positive digits except 7. The factor 61843 comes as a surprise.

O'Shea partitioned the digits of 8549176320 in two different ways to create the following equation:

$$854 + 917 + 632 + 0 = 8 \cdot 5 \cdot 49 + (1 \cdot 7 \cdot 63) + 2 + 0$$

Each side of the equality is 2403.

O'Shea then reversed the alphabetical number to produce 0236719458. By partitioning this number, $0 + 2367 + 19 + 4 + 5 + 8$, he also arrived at the sum 2403.

Two American mathematicians, James Smoak and Thomas J. Osler, in their paper "A Magic Trick from Fibonacci,"[12] report an amazing curiosity. Consider the fraction 100/89. It equals 1.12359550561 . . . Note that the first five digits are the first five Fibonacci numbers.

Add two zeros to the fraction's numerator, then add a 9 to the front and to the end of the denominator to produce the fraction 10000/9899. It equals:

$$1.0102030508132134559046368 \ldots$$

Note that the first 1, followed by the next nine *pairs* of digits, are the first *ten* Fibonacci numbers!

---

*If you can discover other strange properties of 8549176320, send them to me in care of Hill and Wang. Here is one I stumbled across. Divide the alphabetical number by 2718, the first four digits of *e*, and you get a number starting with 314, the first three digits of pi! I also found that 123456789, when divided five times by 5, yields five quotients that each contain all nine positive digits, two also containing a zero.

The authors prove that this procedure, continued forever, will generate *all* Fibonacci numbers! Each step increases the number of Fibonacci numbers by five. Thus, the fractions 1000000/998999, taking decimal numbers in triplets, gives the first *fifteen* Fibonacci numbers! The next fraction produces the first twenty Fibonacci numbers, the next gives the first twenty-five Fibonacci numbers, and so on to infinity!

The above curiosity appears as exercise G43 in *Concrete Mathematics* by Graham, Knuth, and Patashnik,[13] where it is credited to an article by Brooke and Wall in *The Fibonacci Quarterly.*[14] Knuth tells me that similar fractions, such as 1000000/989899 and 1000000000/998998999, generate the Tribonacci numbers!

Few mathematicians, I suspect, realize that the Fibonacci sequence can provide the base for an arithmetic notation. Every positive integer can be expressed in a unique way as the sum of a set of nonconsecutive Fibonacci numbers. And did you know that the twelfth Fibonacci number is $12^2 = 144$? It is the only square Fibonacci number except for 1. The only Fibonacci cubes are 1 and 8. For more such curiosities, see chapter 13 of my *Mathematical Circus.*[15]

Is there a simple way to test whether any given number is a Fibonacci number? There is. An integer n is a Fibonacci number if and only if $5n^2 + 4$ or $5n^2 - 4$ is a square! You might enjoy testing some integers on your calculator. Is 666 a Fibonacci number? No! Is 123? How about 987?

Finally, here is a strange equation that combines the Fibonacci sequence with a sequence of factorials to produce at the limit the value of *e*. Like pi, this famous transcendental number has a way of turning up in all sorts of unexpected ways. The mysterious fraction was sent to me by O'Shea, who says he found it on the Web.

$$e = \cfrac{1+1+\dfrac{2}{2!}+\dfrac{3}{3!}+\dfrac{5}{4!}+\dfrac{8}{5!}+\dfrac{13}{6!}+\dfrac{21}{7!}+\dfrac{34}{8!}+\dfrac{55}{9!}+\ldots}{1+0+\dfrac{1}{2!}+\dfrac{1}{3!}+\dfrac{2}{4!}+\dfrac{3}{5!}+\dfrac{5}{6!}+\dfrac{8}{7!}+\dfrac{13}{8!}+\dfrac{21}{9!}+\ldots}$$

II. L-TROMINO TILING OF MUTILATED CHESSBOARDS

Tiling theory has become a popular topic for today's mathematicians. The following paper appeared in the May 2009 issue of *The College Mathematics Journal*.

INTRODUCTION

Suppose a standard chessboard is "mutilated" by the removal of two diagonally opposite corner cells. Can the remaining 62 squares be tiled with thirty-one dominoes? The answer is no, because the removed squares are the *same* color. Say the color is white. The remaining 62 squares will have an excess of two black cells. Each domino covers one black and one white cell. After 30 are placed, two black cells will remain uncovered. They cannot be adjacent, therefore they can't be covered by a domino. This famous puzzle, solved by a simple parity check, is a simple example of a tiling problem on a mutilated chessboard.

Less well known is the following related problem. Assume the chessboard is mutilated by having two cells removed of *opposite* color from anywhere on the board. Can the remaining 62 squares always be tiled by dominoes? The answer is yes, and there is a lovely proof by Ralph Gomory.[1]

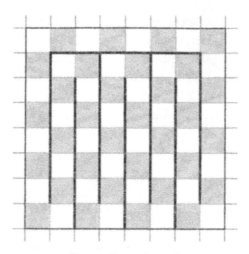

Figure I. Gomory's proof.

Imagine heavy lines drawn on the chessboard as shown in figure 1. They outline a closed path along which the squares lie like beads of alternating color on a necklace. If any two cells of opposite color are taken from the path, it will cut the path into two open-ended segments, or one segment if the removed cells are adjacent. Each segment will consist of an even number of cells of alternating colors, therefore it can be tiled with dominoes. Gomory's clever proof is easily generalized to all square boards with an even number of cells.

If, instead of dominoes, we tile with L-trominoes, also called bent, or V, or right trominoes, then all square boards with a number of cells divisible by 3 can be tiled except for the 3 × 3 board. We will not be concerned with such "whole" boards, but only with mutilated boards with a number of cells that is a multiple of 3 after a single cell has been removed from any spot on the board. We will call such boards *deficient*. In other words, a board of side n is deficient if $n^2 - 1$ is a multiple of 3, i.e., n is *not* a multiple of 3. The sides of such boards form the sequence (*):

$$2, 4, 5, 7, 8, 10, 11, 13, 14, \ldots \; (*)$$

We will call these numbers the *orders* of a board and, from now on, the word *tromino* will mean an L-tromino exclusively.

Our basic question is this: What deficient boards with sides in the sequence (*) can be tiled without gaps or overlaps with L-trominoes after a cell has been taken from anywhere on the board? We will take up these boards roughly in numerical order, culminating with a statement of the complete solution.

POWERS OF 2

Consider the order-2 board first. It obviously is tilable with any cell missing (see figure 2, left). Figure 2, right, shows how the order-4 can be tiled. The 2 × 2 square takes care of a missing cell in each of its four corners. The rest of the board is tiled by taking advantage of what Solomon Golomb named a rep-tile—a tile that can form an enlarged replica of itself. The top left 2 × 2 square rotates to put its missing cell in four places, and the entire order-4 square rotates to carry the missing cell to any of its sixteen places.

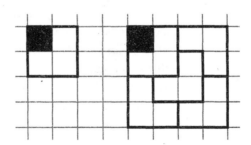

Figure 2. Orders 2 and 4.

In 1953 Golomb, the "father" of polyominoes (he named them and was the first to study them in depth) discovered a beautiful proof by induction that all boards with sides in the doubling sequence 2, 4, 8, 16, . . . could be tiled with trominoes when any cell is missing. The proof was first published in 1938.[2] It is repeated in E. B. Escott's *Open Court* (pp. 27–28).[3] Numerous mathematicians have since included the proof in their books, often without credit to Golomb. Roger Nelsen gives Golomb's proof with a wordless single diagram.[4]

Golomb's famous proof starts with the 2 × 2 case shown on the left of figure 3. This square is placed in the corner of the order-4 as shown at the center of figure 3. The 4 × 4 then goes in the corner of an order-8 (shown on the right) and a tromino placed at the corner of the shaded order-4. We know the dark square can be tiled with any cell missing, and we know the three unshaded quadrants can be tiled with trominoes because each has a missing corner cell. By rotating the board, a missing cell at any spot in the shaded quadrant can be brought to any spot on the order-8 board.

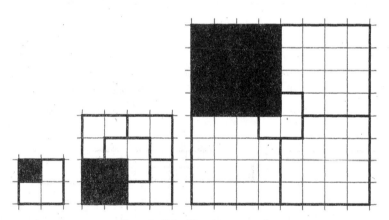

Figure 3. Golomb's induction proof.

ORDERS 5 AND 7

The order-5 board is next, as 5 is the next unsolved number in the sequence (*). It has a neat symmetrical tiling when the center cell is gone, as shown in figure 4, left. I have tiled this board with four 2 × 3 tiles. Each is tilable with two trominoes in two different ways. Using 2 × 3 tiles is a valuable device for solving tromino problems.

When the missing cell is the one shown in black in figure 4, center, the cell above it must be covered by a tromino on either side. In each case, shown here with a tromino above and on the right, this produces two cells (numbered 1 and 2) that cannot be covered with a tile. Indeed, the order-5 square can be tiled only when the missing cell is one of the nine shown in black in figure 4, right. As a pleasant exercise, see if you can tile the board when the missing cell is at a corner.

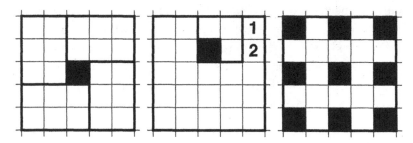

Figure 4. The order-5 square.

The order-7 board is more difficult to analyze. I was unable to find a single diagram that would prove this board tilable, but Golomb sent me his unpublished way of proving tilability with the aid of three diagrams.

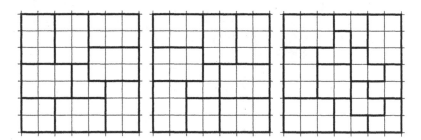

Figure 5. Golomb's proof that order-7 is tilable.

His proof goes like this. Figure 5 shows three tilings of the order-7 board. In each tiling, the 2 × 2 square obviously can be tiled with a tromino so that the missing cell is at any of the four corners. By rotating the three patterns, the missing cell can be placed at any spot on the board.

Somewhat more difficult is to find tilings that maximize the number of 2 × 3 tiles. As a challenge, can you find a tiling of the 7 × 7 board using six 2 × 3 tiles and 4 trominoes (see figure 6)? The solution is unique except for a single reflection. (It appears on p. 123).

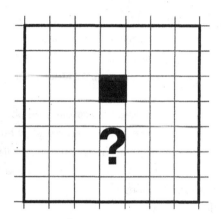

Figure 6. A challenge.

Note in figure 5 that in each pattern the number of free trominoes—trominoes not in any 2 × 3 tile—is always even. This is no coincidence. It led me to the following trivial little law. When a board's order is even, the number of free trominoes in a tiling pattern is odd, and vice versa. When the board's number is odd, the number of free trominoes must be even.

The parity proof is simple. If a board's order is even, after a cell is removed there will be $(n^2 - 1)/3$ trominoes in any tiling, an odd number. Each 2 × 3 tile contains two trominoes, so the total number of trominoes in 2 × 3 tiles will be even. Subtract this number from the odd total of trominoes and you get an odd number of trominoes not in any 2 × 3 tiles.

Suppose the board's order is odd. After a cell is removed there will remain an even number of cells. Subtracting the even number of trominoes in the 2 × 3 tiles leaves an even number of trominoes not in a 2 × 3 tile.

BEYOND 7

Golomb's induction proof can be applied to an infinity of other doubling sequences. In particular, now that we have tiled the 7 × 7 board, we can tile boards of size $n \times n$ where n is of the form 2^k 7. For example, consider the order-14 board. Divide it into quadrants with a shaded order-7 board in the top left corner, and attach a tromino to its lower right corner as before. Because the 7-board is tilable, the proof for order-14 follows, and of course leads by induction to proofs for orders 28, 56, 112, . . .

A similar proof for the order-10 board can't be obtained by placing an order-5 in the corner because order-5 is not tilable, but we can handle it in a slightly different way. Put in the top left corner an order-8, which we know is tilable. The remaining area forms a path of width 2 along the bottom and right sides of the large square (see figure 7). By rotations and reflections, each missing cell in the order-8 can be transferred to any cell on the board. This leads to proofs for orders 20, 40, 80, and so on. A similar proof for order-11 has an order-7 square in the corner, and a path of width 4 along the bottom and side. It leads by induction to solutions for orders 22, 44, 88, . . . Clearly this technique

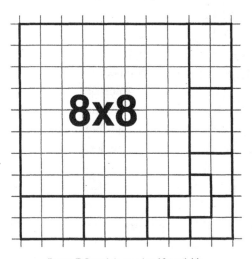

Figure 7. Proof that order-10 is tilable.

provides an infinity of doubling sequences for tilable boards. Simply put in the top left corner of any board a tilable board with a side equal to or smaller than the larger board. If you can tile the path that the smaller board leaves at the bottom and side, then the larger board is tilable.

Boards with sides that are primes are usually the hardest to tile. Order-17 is solved by a corner square of side 13 and a path of width 4. Order-19 is solved by a corner square of order-14, in turn based on order-7, and a path of width 5. (See figure 8.)

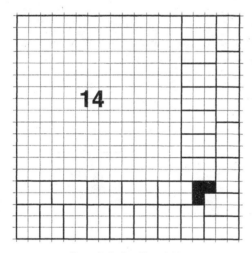

Figure 8. Order-19 is tilable.

THE COMPLETE RESULT

By working with these patterns I came close, but not close enough, to finding an induction proof that all deficient squares are tilable except for order-5. A proof was finally obtained by I. Ping Chu and Richard Johnsonbaugh.[5]

Chu and Johnsonbaugh not only took care of all deficient squares, but also all deficient rectangles! Their induction proof is too technical to repeat here. To summarize, they showed tilability for all $m \times n$ rectangles (including squares when $m = n$) which have a number of cells that is a multiple of 3 after a cell is removed. Such boards are tilable if and only if all of the following are true:

1. m is equal to or greater than 2.
2. n is equal to or greater than m.
3. If m is 2, n must be 2.
4. m is not 5.

A 4 × 7 rectangle is the smallest deficient rectangle, not a square, that is tilable with L-trominoes. As another exercise, see how long it takes you to tile it when the missing cell is at a corner, and there are two 2 × 3 tiles.

Christopher Jensen, in a 1995 paper, showed that if *two* cells are taken from a corner of any board, as shown in figure 9, the board obviously cannot be tiled with trominoes. However, if none of these five cases is allowed, a $3m - 1$ by $3n + 1$ board with any two cells missing can be tiled if and only if $n = 1$ or m and n are each equal to or greater than 3.[7]

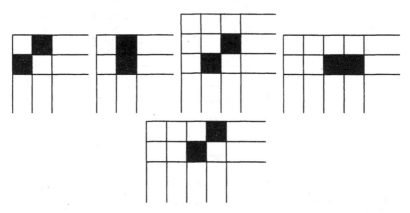

Figure 9. Impossible tiling patterns when *two* cells are missing at a corner.

A FINAL WORD

Kate Jones, who founded and runs Kadon Enterprises, a firm that makes and sells handsome mechanical puzzles, games, and other recreational math items, has on the market a game called Vee-21.[6] The Vee is for V-trominoes, and 21 for the 21 tromino tiles in the set. The trominoes are brightly colored, and there is an order-8 board on which to place them. The basic task is to put a monomino (order-1 tile) at any spot on the board, then cover the remaining 63 cells with the trominoes, thus solving an order-8 board. A forty-page brochure comes with the set. It contains a short article on "The Deficient Checkerboard" by Norton Starr, and pictures of rectangular fields that offer other challenges.

Our final tiling (see figure 10) is a beautiful, symmetric tiling of the standard chessboard.

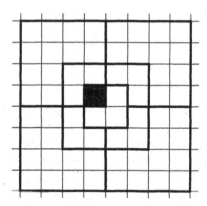

Figure 10. An order-8 tiling with no 2 × 3 tiles and 5 rep-tiles.

Solution to the reader challenge problem (see page 119):

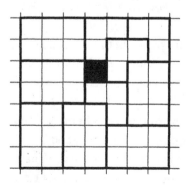

12. IS REUBEN HERSH "OUT THERE"?

Reuben Hersh belongs to a small group of mathematicians convinced that mathematics has no reality apart from human cultures. I am an unashamed Platonist who prefers a language that assumes that if all sentient beings in the universe disappeared, there would remain a sense in which mathematical objects and theorems would continue to exist even though there would be no one around to write or talk about them. Huge prime numbers would continue to be prime even if no one had proved them prime. As Bertrand Russell once wrote, two plus two is four even in the interior of the sun.

For an earlier paper on the same topic, see "A Defense of Platonic Realism," chapter 9 in my *Jinn from Hyperspace* (Amherst, NY: Prometheus Books, 2008).

Brian Davies, in his paper "Let Platonism Die" (*IMS Newsletter*, June 2007), defines mathematical Platonism as the belief that mathematical entities exist "in a mathematical realm outside the confines of space and time." This is not what I, or, I think, most mathematical Platonists believe. Aristotle, a mathematical realist, grabbed Plato's universals (redness, cowness, twoness, and so on) from a transcendental realm, and attached them to objects in space and time. The geometrical shape of a vase, for example, is "out there," on the vase, not something floating outside Plato's cave.

Consider pebbles. On the assumption that every pebble is a model of the number 1, obviously all the theorems of arithmetic can be proved by manipulating pebbles. Even in principle you can prove that any integer, no matter how large, is either prime or composite.

Reuben Hersh, my old adversary, in a paper "On Platonism" (*IMS Newsletter*, June 2008) says this:

> My view of Platonism—always referring to the common, everyday Platonism of the typical working mathematician—is that it expresses a correct recognition that there are mathematical facts and entities, that these are not subject to the will or whim of the individual mathematician but are forced on him as objective facts and entities which he must learn about and whose independent existence and qualities he seeks to recognize and discover.

Welcome, Professor Hersh, to the Plato club! All Platonists would agree completely with your remarks. But then Hersh goes on to make an incredible statement: "The fallacy of Platonism is the misinterpretation of this objective reality, putting it outside of human culture and consciousness."

The theorems and objects of mathematics, Hersh continues, like "many other cultural realities" are "external, objective, *from the viewpoint of any individual* [italics his], but internal, historical, socially conditioned *from the viewpoint of the society or that culture as a whole* [italics still his]."

So, Hersh is not a Platonist after all! Does he really think that manipulating pebbles to prove, say, that 17 is a prime is not a process going on out there, unconditioned by a given culture? Of course manipulating pebbles is culturally conditioned in the trivial sense that everything humans do is so conditioned. But that is not the deeper question. The primality of 17, in an obvious way, is out there in the behavior of pebbles in much the same way that the elliptical orbit of Mars is out there, or the spiral form of our galaxy.

Hersh is so addicted to squeezing math inside folkways that in his book *What Is Mathematics, Really?* (Oxford University Press, 1997) he writes, so help me, that 8 plus 5 is not necessarily 13, because some skyscrapers have no thirteenth floor. So, if you go up eight floors in an elevator, then go five more floors, you find yourself on the fourteenth floor. Is Hersh suggesting that in the subculture of some skyscrapers the laws of arithmetic are constantly violated?

Need I point out that since geometry was arithmetized by Descartes, it too can, in principle, be modeled with pebbles? Indeed, the universe is saturated with models of nearly all of mathematics. Even a topologist can prove that bisecting a Klein bottle produces two Möbius bands of opposite handedness by making a crude model out of an envelope, then cutting it in half.[1]

Complex numbers and derivatives may not have material models, but they also are embedded throughout the universe. Newton and Leibniz in an obvious sense invented calculus, but in another sense they discovered fundamental ways in which the universe behaves. The Mandelbrot set is not outside space and time. It exists on computer screens. Does an antirealist believe that a mathematician exploring properties of the Mandelbrot set is really exploring a structure inside his brain because his eyes and brain are seeing the screen, or that he is exploring part of his culture because his culture built the computer?

Such statements involve the same kind of distortion of language as claiming that astronomers are not studying patterns "out there" because telescopes are part of culture—not to mention all of astronomy as well. This is not far from insisting that the universe exists because human cultures observe it, rather than that we exist because the universe fabricated us.

Cantor's alephs may not be out there, but who knows? They could be hiding somewhere in the cosmos. Like physicists, mathematicians often discover things by investigating material models. Frank Morley, to consider a classic instance, discovered "Morley's theorem" by investigating the angles of paper models of arbitrary triangles—models as much "out there" as stones and stars. In no way can it be said that Morley *invented* his theorem, or found it inside his skull or as part of his culture.

In his paper Hersh correctly calls me a theist. He adds that I believe in the efficacy of prayer. Hersh, an atheist, considers this an insult. Well, it all depends on what "efficacy" means. I don't believe that if someone prays for a football team to win a game, or for a loved one to have a remission of cancer, that God will insert a hand into the universe and change it accordingly. I suppose it is possible for God to alter probabilities on the quantum level, now a popular conjecture by theists, but I'm inclined to doubt it.

I *do* think prayers for forgiveness are justified, or prayers for the wisdom to make right decisions. Gilbert Chesterton somewhere says that it is a sad day for atheists when something wonderful happens to them and they have nobody to thank.

Hersh also writes that I once accused him of Stalinism. I can't imagine how I could have done such a thing. If I did, I apologize. Perhaps I once reminded him of that harrowing scene in Orwell's *1984* in which an official manages to torture a prisoner into believing that when two fingers are added to two fingers a fifth finger appears.

Hersh also claims that I once accused him of being a solipsist. Again, I'm at a loss to know what he has in mind. It is possible that I described his antirealism as a vague kind of social or collective solipsism. Hersh is a great admirer of a paper by the anthropologist Leslie White titled "The Locus of Mathematical Reality." The locus, White argues, is not in the outside world but in human cultures. Mathematical theorems are similar to such things as traffic regulations, fashions in clothes, art, music, and so on.

This of course isn't solipsism in the usual sense. No one outside a mental hospital is a true solipsist. But the antirealism of White and Hersh is tinged with social solipsism in the sense that if human culture were to disappear, all of mathematics would also vanish. True, the universe would persist, but there would be nobody around to do mathematics unless there are mathematicians on other planets. Presumably Hersh would agree that what we call mathematical structures and events would still exist, but if there were no sentient beings to study them there would be nothing in the universe deserving the name of mathematics.

Here once more the question arises of what is the best, the least confusing language to use. I think it best to say that if all sentient creatures vanished, 2 plus 2 would still be 4, the circumference of the moon's disk divided by its diameter would still be close to pi, and Euclidian triangles would still have interior angles that added to a straight angle. I assume that Hersh would prefer to say that none of these assertions would still hold because there would be no cultures around in which such statements could be made. To think otherwise would, God forbid, turn Hersh into a Platonist.

Like Paul Dirac and thousands of other eminent mathematicians, I believe there is a God who is a superb mathematician with a knowledge of math enormously superior to ours. Whether infinite or not, how could I know? God surely does not know the last decimal digit of pi, because there *is* no last digit. Even if I were an atheist I would still consider it monstrous hubris to suppose that mathematics has no reality apart from the little minds of intelligent monkeys.

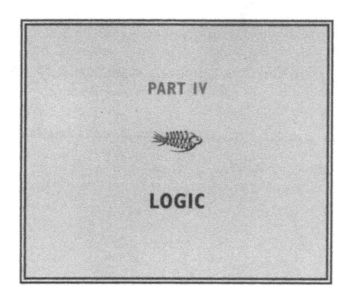

PART IV

LOGIC

13. THE EXPLOSION OF BLABBAGE'S ORACLE

Whenever a prediction is part of the event being predicted, logical paradoxes can arise. I have written about such paradoxes in several places. This chapter was first published in *Isaac Asimov's Science Fiction Magazine* (August 1979). My earlier version of the paradox, in the form of a bar bet, can be found in *Ibidem* (a Canadian magic periodical), March 1961, and in chapter 11 of my *New Mathematical Diversions from "Scientific American"* (New York: Simon & Schuster, 1966). A much more mystifying prediction paradox is the topic of the next chapter.

Professor Charles Blabbage, England's top expert on artificial intelligence, finally completed his construction of ORACLE, an acronym for Omniscient Rational Advance Calculator of Local Events. The computer was so powerful that it could (Blabbage maintained) predict with 100 percent accuracy any event in the laboratory within a period of one hour and inside a radius of ten meters from the computer's console.

This is how it operated. One could describe to ORACLE any event that would or would not occur during the next hour and within the specified radius. If the computer predicted that the event would take place it turned on a green light for yes. If it predicted the event would not take place it turned on a red light for no.

It was necessary, Professor Blabbage made clear, that the two lights be concealed until the hour was up. Otherwise anyone could easily render a prediction wrong by doing something to falsify it. For example, suppose the computer predicted yes for: "A cockroach will crawl across the west wall of the lab." If someone saw the green light he or she could stand guard by the wall to make sure the event did not occur.

Blabbage's assistant was Dr. Ada Loveface, an attractive young redhead with a doctorate in modern logic and set theory. On the day before Blabbage was to demonstrate ORACLE's powers for a group of distinguished visiting computer scientists, military moguls, and government officials, Dr. Loveface approached him and said: "I regret having to tell you this, Professor, but I've just proved that ORACLE can't possibly succeed in all cases. I can describe an event that will or will not take place in the lab, within the hour and inside the ten-meter radius, of such a nature that the computer will find it logically impossible to predict whether it will or won't happen."

Blabbage refused to believe Ada until she told him what the event was. Her remarks were so shattering, he collapsed in a faint and had to be taken to a hospital.

What event did Dr. Loveface think of?

Dr. Loveface thought of the following event: "ORACLE will make its next prediction by turning on its red light."

This would force the computer into a logical contradiction. If it turned on the red light for no, the prediction would be wrong because the red light did in fact go on. If it turned on the green light for yes, this too would be wrong because the green light went on, not the red.

While Professor Blabbage was recuperating, Dr. Loveface actually gave the event to ORACLE and requested its prediction. The computer's circuits went into a yes–no loop, producing a humming sound that grew steadily louder until suddenly the entire computer exploded, completely destroying Blabbage's lifework.

There are many variations of this basic paradox which show that under certain conditions predictions of the future are impossible in principle. Can you think of an equivalent version of the computer paradox so simple that you can inflict it on a friend by speaking fewer than fifteen words?

I do not know who first thought of the red-and-green-light version of the computer prediction paradox. It was the basis for a variation I introduced as a betting game in a *Scientific American* column that became chapter 11 of my *New Mathematical Diversions*. For a prediction paradox more difficult to resolve than this one, see the next chapter of this book.

Charles Blabbage is an obvious play on Charles Babbage, the British pioneer of computers that can be programmed. Babbage's good friend and disciple, Ada Augusta, the beautiful young countess of Lovelace, was wealthy, witty, intelligent, a good mathematician, and the only legitimate child of the poet Lord Byron. She was the first to say that computers do only what they are told to do. The character of Ada, in Nabokov's novel *Ada*, is partly based on Lady Lovelace.

If you want to know more about this remarkable pair, see *Charles Babbage and His Calculating Engines* by Philip and Emily Morrison; *Ada, Countess of Lovelace* by Doris Langley Moore; and "Byron's Daughter" by B. H. Neumann in the *Mathematical Gazette* 57, June 1973, pp. 94–97.

14. THE ERASING OF PHILBERT THE FUDGER

The logic paradox presented in the following chapter was called the paradox of the unexpected examination when it was first introduced by D. J. O'Connor in his paper "Pragmatic Paradoxes" (*Mind* 57, July 1948). It was later called the paradox of the unexpected hanging. For an in-depth analysis, see the first chapter of my book *The Unexpected Hanging* (New York: Simon & Schuster, 1969). I give there a lengthy bibliography, far from complete because so many papers on the paradox have since been published. The whimsical version here first ran in *Isaac Asimov's Science Fiction Magazine*, November 1969.

By the mid-twenty-third century capital punishment has been replaced throughout most of the civilized world by a punishment called "erasure." The criminal's head is placed inside an electronic machine called the "oblivion box." It takes only a few minutes to expunge from the brain all memories of events experienced after the first six months of life. This, of course, returns the criminal to babyhood. It has long been established that no one is born with criminal tendencies—all are acquired by experience. Over a period of years the erased "baby" slowly develops into a new adult. Because erasure turns a person into a different personality, with no memory of his or her former self, the punishment is feared almost as much as execution.

Another radical change in the administration of justice is the replacement of all judges, and some lawyers, by robots. Laws have become so numerous and complicated that only computers can remember all the details. Robot judges are carefully programmed to make only wise and logical decisions. It is impossible for a robot judge to lie. If a circuit in his brain malfunctions, and he makes any statement that is false, his pronouncements are declared null and void and a new trial is scheduled.

One of the most heinous crimes in the twenty-third century, on a par with murder and rape, is the crime of "fudging." This means a falsification of data in a scientific experiment. Such an enormous respect has developed for the sanctity of the scientific method that anyone declared guilty of fudging is automatically sentenced to erasure.

Philbert X1729B was arrested for having fudged the data in a tooth-decay experiment he had supervised at the Oral Roberts Dental Research Laboratory in Tulsa, Oklahoma. Philbert could not afford a robot lawyer. His human lawyer, who was not very skillful, lost the case. At Philbert's sentencing the robot judge said:

"You will be erased at 3 p.m. on one of the six days of next week, starting with Monday. You'll be informed of the day at 10 a.m. on the day of the erasure."

"But, Judge," said Philbert, "can't you tell me the day now?"

"No. The date has not yet been determined. I can assure you, however, that it will be Monday, Tuesday, Wednesday, Thursday, Friday, or Saturday of next week. You won't know what day it is until we inform you on the morning of the erasure day."

"Thank you, your honor."

It was the last sentencing of the day, so the judge pressed a button under his left armpit to turn himself off until the court opened the next morning.

Sitting in his cell, Philbert began to think about what the judge had said. Suddenly he leaped to his feet with a yelp of joy. There was no way he could be erased without making the judge out to be a liar! This would guarantee him a new trial. Maybe his wife and friends would be able to raise enough money for a good robot lawyer.

What was Philbert's reasoning?

Philbert reasoned as follows:

"Suppose my erasure day is Saturday. No one will tell me Friday morning that Saturday is the day, therefore on Friday afternoon I will *know for certain* that the day is Saturday. But the judge told me I would *not* know the day until the morning of the day itself. Therefore I *can't* be erased on Saturday without making the judge a liar.

"Consider Friday. It too is ruled out. Since Saturday cannot possibly be the day I stick my head in the oblivion box, if I'm not told the day by Thursday noon, I will know that the day is Friday. Why? Because only Friday and Saturday remain. It can't be Saturday, hence it must be Friday. But if I know on Thursday that it is Friday, the judge again will have uttered a falsehood.

"So Friday and Saturday are out. Consider Thursday. It too is eliminated by the same reasoning! After twelve o'clock on Wednesday, if I've not been told the day, I will know it is Thursday because it can't be Friday or Saturday. The same reasoning applies to Wednesday, Tuesday, and Monday. No matter what day is picked, I'll know the date by the afternoon of the previous day. In each case it will make the judge a liar and allow me a new trial."

Philbert's reasoning seems impeccable, yet there is a fatal flaw in his logic. It is not so easy to pinpoint exactly where the flaw lies, but it *is* easy to prove that Philbert's reasoning can't be correct. How?

On Thursday morning Philbert was told he would be erased that afternoon. Since Philbert had no way of knowing it would be Thursday, this news came to him as a total surprise. His erasure took place on Thursday. Everything the judge said proved to be accurate.

POSTSCRIPT

This is one of the most notorious of the prediction paradoxes of modern philosophy. A fuller discussion of the paradox, and a listing of twenty-three papers on it, is found in my book, *The Unexpected Hanging*, chapter 1. Since the book's publication in 1969, more than a dozen new papers have been published. They are listed in an extensive bibliography accompanying "Expecting the Unexpected," by Maya Bar-Hillel and Avishai Margalit.

PART V

LITERATURE

15. THE WONDERFUL WIZARD OF OZ

I learned to read by looking over my mother's shoulder at the pages of the book she was reading aloud. The book was Lyman Frank Baum's *The Wonderful Wizard of Oz*. A few years later I persuaded my parents to buy all of Baum's other Oz books, as well as his many fantasies about enchanted lands other than Oz.

As an adult I wrote a great deal about Baum and his books, including a biography of Baum that ran in two issues of Anthony Boucher's magazine *Fantasy and Science Fiction*. It was reprinted in Michigan State University Press's edition of *The Wonderful Wizard of Oz and Who He Was* (1957).

The following chapter was originally my introduction to a Dover paperback edition of *The Wizard of Oz*. The book included full-color reproductions of William Wallace Denslow's illustrations for the book's first edition.

In the fall of 1900, when *The Wonderful Wizard of Oz* first went on sale, Lyman Frank Baum was forty-four. His life had been a restless one: newspaper reporter in New York, manager of a chain of theaters, playwright and actor, owner of a variety store in Aberdeen, South Dakota, editor and publisher of Aberdeen's weekly newspaper, road salesman of china and glassware, founder (in Chicago) of a national association of window trimmers and editor of their official organ, *The Show Window*.

It was in Chicago that Baum began writing books for children. *Mother Goose in Prose*, published by Way and Williams in 1897, introduced in its final story a little farm girl named Dorothy. Two years later, the Chicago firm of George M. Hill brought out *Father Goose: His Book*, a collection of nonsense verse written by Baum and illustrated by his friend William Wallace Denslow, a Chicago newspaper artist. It sold surprisingly well and Baum began to work on children's books in earnest. The following year, when *The Wonderful Wizard of Oz* was issued by George Hill, it became an instant success. In a letter written many years later, Mrs. Baum recalls asking her husband for a hundred dollars during the Christmas season of 1900, and her astonishment when he produced a royalty check for $2,500. Other publishers rushed books into print that were obvious imitations of *The Wizard*, both in style of writing and book design.

Baum and Denslow collaborated on another delightful fairy tale, *Dot and Tot of Merryland*; then they had a falling out, and John Rea Neill, a young Philadelphia artist, became the illustrator for all Baum's later Oz books. *The Wizard* has since been illustrated by a dozen different hands, but none have had the quaint, whimsical touch that has linked Denslow's pictures almost as firmly to the book as Tenniel's pictures are linked to Lewis Carroll's *Alice in Wonderland*.

In *The Show Window*, October 15, 1900, Baum wrote his valedictory editorial. "The generous reception by the American people of my books for children, during the past two years, has resulted in such constant demands upon my time that I find it necessary to devote my entire attention, hereafter, to this class of work." And that is just what he did. In the remaining nineteen years of his life he wrote more than sixty books for children, many appearing under pseudonyms. About half of these books were fantasies, of which fourteen (including *The Wizard*) were in the Oz series. After Baum's death in 1919, nineteen more Oz books were written by Ruth Plumly Thompson, a Philadelphia writer. Three Oz books were written by Neill, two by Jack Snow, one by Rachel Cosgrove, and one by Baum's son, Colonel Frank Joslyn Baum, making a total of forty Oz books. The Colonel's book, now a scarce collector's item, was titled *The Laughing Dragon of Oz*. It was issued by the Whitman Publishing Company in 1934, and sold in the dime stores for ten cents.

Although some of Baum's other fantasies are, in my opinion, better written than any of his Oz books, and some of his Oz books better written than *The Wizard*, it is *The Wizard* that has become this country's greatest, best-loved fairy tale. It has never been out of print, and so many different editions have been published, in the United States and abroad, that no one knows how many millions of copies have been sold. Some of these editions take strange liberties with the text. For example, in chapter 14, almost all recent printings contain references to scarlet fields and scarlet flowers, though Dorothy and her friends are in Winkie country, where yellow is the dominant color. This was not an oversight on the part of "The Royal Historian of Oz," as Baum liked to call himself. His original text speaks only of buttercups and daisies, yellow flowers and yellow fields. (This odd corruption of the canon was discovered by Dick Martin, a Chicago artist and leading authority in Oziana.)

In 1902, Baum was persuaded to write the book and lyrics for a musical comedy show based on *The Wizard*. It opened in Chicago and was such a smash hit that it moved to New York in 1903 and for eighteen months played to capacity houses on Broadway. The show's plot departed greatly from the original story, and introduced such characters as a lady lunatic, a boy poet (with whom Dorothy fell in love), and Pastoria, a streetcar motorman from Topeka. Since it obviously was not possible for an actor to disguise himself as a little black dog, Baum changed Dorothy's pet to a large spotted calf named Imogene! A former circus acrobat, Fred Stone, became a famous comedian almost overnight for his portrayal of the Scarecrow.

Several motion picture versions of *The Wizard* have been made. The first two were mediocre silent films: a one-reeler produced by Selig Pictures in 1910, and a seven-reeler, featuring the comedian Larry Semon as the Scarecrow, issued by Chadwick Pictures in 1925. It is interesting to learn that in the Chadwick film the role of the Tin Woodman was taken by Oliver Hardy, fat man of the Laurel and Hardy comedy team. The last and greatest filming was, of course, the 1939

Metro-Goldwyn-Mayer color extravaganza starring young Judy Garland as a singing Dorothy. Ray Bolger danced the part of the Scarecrow, Jack Haley clanked his way through the role of the Tin Woodman, and Bert Lahr was an amusing, magnificently cowardly, Cowardly Lion.

It is not hard to understand why *The Wizard* is read by today's boys and girls with the same wide-eyed wonder with which it was read in 1900. It is a superb, skillfully written fantasy, blazing with color and excitement, rich in humor and quiet wisdom. A child may not be aware of the story's satirical thrusts and higher levels of meaning, but they are there, and they are one of the reasons why *The Wizard* has become the classic that it is. Was T. S. Eliot thinking vaguely (among other things) of the Tin Woodman and the Scarecrow when he wrote, "We are the hollow men. We are the stuffed men"? Are the respected Wizards of our Emerald Cities really wizards, or just amiable circus humbugs who keep us supplied with colored glasses that make life seem greener than it really is?

We are all little children walking down a road of yellow brick in a crazy, outlandish, Ozzy sort of world. We know that wisdom, love, and courage are essential virtues, but like Dorothy we cannot decide whether it is best to seek for better brains (our electronic computers grow more powerful every year!) or for kinder, more loving hearts.

16. THE LIFE AND ADVENTURES OF SANTA CLAUS

Very few Americans, I suspect, are aware that L. Frank Baum wrote fourteen Oz books, and almost as many other fantasies that take place in Ev and other magic lands outside of Oz. One of his best non-Oz books is his life of Santa Claus. The locale is the Forest of Burzee, just beyond the Deadly Desert at the southern border of Oz.

The chapter here reprints my introduction to a Dover paperback edition (New York, 1976) of *The Life and Adventures of Santa Claus*.

In all this world there is nothing so beautiful as a happy child.
—L. Frank Baum's Santa Claus

Mythology in the United States, as in ancient Rome, is largely derivative. We do have a few native legends (Johnny Appleseed and Paul Bunyan, to name two) and some tall tales about George Washington and other early dignitaries, but our great myths are borrowed from the Judeo-Christian traditions of Western Europe.

With one exception! There is a single towering personality, as immortal as Pan, whose character has almost entirely been shaped by the American imagination. More surprisingly, as we shall see, his distinguishing features were mainly fashioned by residents of Manhattan. I am speaking, of course, of Santa Claus.

Let us not confuse Santa with his pale predecessors and curious counterparts. Saint Nicholas, the fourth-century bishop of Myra (in Asia Minor), is usually portrayed as tall and lean. He was much venerated throughout the Middle Ages, especially in Russia and Greece; in those two lands he became a national patron saint. By the twelfth century, his feast day (December 6) had become an official church holiday throughout Europe, but after the Reformation he fell into disrepute in Protestant countries, and his veneration faded even in Catholic regions. It was not long until the feast of St. Nicholas had merged with December 25, which for a thousand years had been the official day for celebrating the birth of Jesus.

Holland was the only Protestant country where St. Nicholas survived, and there he became a gift giver resembling Santa Claus. For six hundred years, on the eve of St. Nicholas, Dutch children have been putting their shoes by the fireplace, along with some food for the saint's horse. During the night, Sinterklaas and his Moorish assistant Zwarte Piet (Black Peter) arrive by ship from, of all places, Spain. The saint mounts a white horse that gallops through the air to carry him from roof to roof. Black Peter somehow follows. The Moor pops down chimneys to leave gifts, but Sinterklaas, not wishing to soil his white robe and red cassock, does no more than drop candy down the chimney and into the shoes.

The histories of legendary gift bringers in other lands who make midwinter visits to homes are colorful and confusing. There is Father Christmas in England, an old white-bearded gent who, according to a splendid essay by G. K. Chesterton, has been dying in England ever since the Middle Ages. His counterpart in Paris and French Canada is Père Noël. Germany's Kriss Kringle, actually *Christkindl* or Christ Child (curiously, in the United States, in the mid-nineteenth century, "Kriss Kringle" became a name for Santa Claus), is not the infant Jesus but a kind of fairy who brings the gifts.

In Italy on Twelfth Night (January 5) it is the good witch Befana who slides down the chimney on her broom to fill shoes and stockings with candy and toys. The figure of Befana can be traced back to ancient Roman belief, but there is also a Christian legend that she was sweeping her house when the Three Wise Men came by and asked her to accompany them to Bethlehem. Befana was too busy to leave. Later she regretted her decision and has been wandering around the world ever since under a curse like that of the Wandering Jew. Once a year she searches the houses, peering at the faces of sleeping infants, hoping to find the Christ Child. Her counterpart in pre-Revolutionary Russia was the evil Baboushka (old peasant woman), who intentionally misdirected the Wise Men. In Spain, on Twelfth Night, the Wise Men themselves arrive by horse or camel to leave gifts.

The American Santa Claus is well known in Canada, and his influence has spread to more distant regions. In the Scandinavian nations a fat, white-bearded old fellow arrives by sled, carrying a pack of toys. Reindeer draw his sleigh in Norway and Finland. The Soviet Union officially abolished Christmas (January 7 in the Russian calendar), on which St. Nicholas used to put gifts around a tinseled fir tree. The tree was decorated on New Year's Eve and it was Grandfather Frost who brought the toys. Grandfather Frost was fat, had a white beard, wore a red suit, and looked just like Santa. In Tokyo, Christmas Eve is a major Saturnalia. "Jingoru Beru" (Jingle Bells) can be heard everywhere, people go around saying "Meri Kurisumasu," and fashionable department stores dress their pretty elevator starters in miniskirted Santa suits.

But enough about Santa's imitators. In Colonial America the first Dutch settlers of New Amsterdam (New York) naturally brought St. Nicholas with them. Washington Irving, in his mock-serious *History of New York* (1809), was the first to write about the Americanized Saint Nick, and scholars are still disputing how much of what Irving said is authentic. Exactly what the saint looked like is not made clear, but we are told that Dutch children would hang their stockings by the fire on St. Nicholas' Eve, and the saint would come "riding over the tops of trees" in a "wagon" to send toys and candy rattling down the chimney.

Now comes the most amazing part of our story. Dr. Clement Clarke Moore was a professor of Greek and Hebrew at an Episcopalian seminary he had helped establish near his home in what is now called the

Chelsea section of Manhattan. One snowy Christmas Eve, in 1822, Dr. Moore recited to his children a poem he had written for them. Eventually it found its way to the editor of the Troy (New York) *Sentinel*, who published it anonymously on December 23, 1823, as "An Account of a Visit from St. Nicholas."

The poem was widely reprinted around the country, but Moore did not acknowledge authorship until it appeared in a collection of New York verse in 1837. Later he included it in a book of his own poems. None of the serious poems in this 1844 book is now remembered, but the "Visit," which Moore dashed off so carelessly, has become the greatest Christmas poem in English. It has been reprinted thousands of times and illustrated by hundreds of artists, and has spawned scores of parodies and sequels, of which "Rudolph the Red-Nosed Reindeer" is the latest and best.

The idea of Santa himself coming down the chimney seems to have been original with Moore. For this it was necessary to make St. Nick small and elflike. It was probably also Moore who gave Santa his twinkling eyes, rosy cheeks, cherry-red nose, and pack of toys, but the pipe and the gesture of finger on nose came from Irving. "And when St. Nicholas had smoked his pipe," Irving wrote, "he twisted it in his hatband, and laying his finger beside his nose gave the astonished Van Kortlandt a very significant look; then mounting his wagon he returned over the treetops and disappeared."

"That sleigh drawn by reindeer was pure inspiration!" exclaims Burton E. Stevenson in his *Famous Single Poems*, but we now know that this, too, was not original with Moore. In 1821, a year before Moore wrote his jingle, a tiny book called *The Children's Friend: A New Year's Present, to Little Ones from Five to Twelve* was published in New York. It is exceedingly rare, but you will find its eight pages reproduced in color in *American Heritage*, volume 12, December 1960. Each page has a quatrain of verse under a picture. The first stanza:

Old Santeclaus with much delight
His reindeer drives the frosty night

O'er chimney tops, and track of snow,
To bring his yearly gifts to you.

The illustration shows Santa in a sleigh drawn by one reindeer. We do not know if Moore saw this book. Nor do we know if the sleigh and reindeer were created by its anonymous author or whether they had already become part of Manhattan's Christmas folklore.

Moore's poem does not disclose where St. Nick lives, but the reindeer and the saint "dressed all in fur from his head to his foot" suggest polar regions. From this it was but another step to furnish Santa with a factory near the North Pole and plenty of helpers, perhaps even a wife. And it was not difficult for the Dutch "Sinterklaas" to become "Santa Claus."

The second major contributor to Clausology was the German-born newspaper cartoonist Thomas Nast. He, too, was a New Yorker, best known today for his attacks on the Tammany Tiger, a beast he created along with the Elephant and Donkey as symbols of the two major political parties. Nast's first drawing of Santa was in 1863. It was followed by scores of others, mostly in the 1880's for Christmas issues of *Harper's Weekly*. Nast made two significant changes in Moore's image of the saint. He sometimes enlarged Santa to a fat man of normal height, and he replaced the fur with a red satin suit trimmed with white ermine. The pointed hat, cowhide boots, and wide black belt were other Nast touches.

It would be strange indeed if L. Frank Baum, our greatest author of juvenile fantasy, had neglected the greatest of our native myths. In his first book of fantasy, *Mother Goose in Prose* (1897), each chapter is based on a familiar Mother Goose rhyme. A few years later he conceived of writing a full-length novel that would do much the same thing for the Santa myth. The George M. Hill Company, the Chicago house that published *The Wonderful Wizard of Oz*, planned to issue Baum's *Life and Adventures of Santa Claus* in 1902, but the firm went bankrupt and the contract went to Bowen-Merrill in Indianapolis.

The first printing of Bowen-Merrill's first edition, in 1902, had a red cloth cover stamped in black, green, tan, and white. The quickest way to identify it is by the section headings, which are worded simply "Book First," "Book Second," and "Book Third." The illustrations are twenty color inserts by Mary Cowles Clark, six in full color, fourteen in red and black. The book received more than a hundred newspaper reviews, almost all of them praising both the story and the pictures.

Clark was a native of Syracuse, where Baum himself had once lived. Sometime before 1902, when Baum was there to visit relatives, he met Clark and was favorably impressed by her artistic talents. Apparently it was he who asked her to illustrate his book. An undated contemporary clipping from a Syracuse paper, preserved in one of Baum's scrapbooks, reports that Clark is completing her illustrations for the book and that she designed its cover. A much later interview with Clark, in the Syracuse *Post Standard*, November 10, 1941, is about an exhibition of her watercolors and needlework. The reporter tells us that she is "the successful illustrator of many children's books," but the only two books mentioned are Baum's and a cookbook by Linda Hull Larned.

In the second printing (1902) of *The Life and Adventures of Santa Claus* the section headings are changed to "Youth," "Manhood," and "Old Age." Eight color plates are left out, but numerous marginal illustrations are added. The book you now hold is a reprint of this second printing. All twenty of the original color plates, however, are here retained in black and white. (For detailed points on the various editions, see Dick Martin's article in *The Baum Bugle*, Christmas 1967, p. 17.)

In Baum's day, little or nothing was generally known about Santa's great European antiquity; the American mythology presented him as an old man without an early life history. It is this enormous gap that Baum now proposed to fill by constructing an elaborate background mythology. The only part of the American myth that he did not accept was Santa's residence at the North Pole. The locales in Baum are the great Forest of Burzee, just across the Deadly Desert at the southern border of Oz, and the Laughing Valley of Hohaho, which adjoins Burzee on the east.

In Baum's mythology, God is called the Supreme Master. Beneath him is a trinity of lesser deities: Ak the Master Woodsman, Kern the Master Husbandman, and Bo the Master Mariner. Under the grizzle-bearded Ak are the merry Ryls, who watch over the flowers of earth, the crooked Knooks who watch over the beasts, the Wood-Nymphs who watch over the trees, and the Fairies who watch over human beings like the guardian angels of Christianity.

Zurline, the golden-haired queen of the Wood-Nymphs, should not be confused with Lurline, leader of the fairy band that enchanted Oz. Necile is the lovely Wood-Nymph who, hundreds of years ago, adopted a baby that had been abandoned just outside the Forest of Burzee, and called him Claus. In the language of the Wood-Nymphs this means "little one." Zurline suggested the improvement Neclaus, meaning "Necile's little one." (Baum tells in a footnote how this became corrupted to Nicholas, thus explaining the "false" association of Claus with the Catholic saint.) The little one grows up with a feeling of love and pity for all mortal children on whom, as on himself, the "doom of mankind" has not yet fallen.

"Then why," Claus asks Ak, "if man must perish, is he born?"

Everything has its purpose, explains Ak, and mortals who are "helpful" are "sure to live again." "Even the mortals, after their earth life, enter another existence for all time," Ak says later. It is the only occasion in all of Baum's fantasy where the doctrine of survival after death is explicitly defended.

Convinced that his mission in life is to leave the world better than he found it, Claus leaves Burzee to settle in the Laughing Valley, where, aided by the Knooks, he builds a log house, using only fallen trees because he will not destroy living ones. We learn how he carves his first toy, how the Ryls teach him to paint his carvings, how he discovers the delight of the poor children to whom he gives his wooden animals.

There is no Satan in Baum's mythology, but there are many kinds of evil creatures. Loathsome Gadgols attack the trees. Shiegra, a lioness, would have eaten the baby Claus had not Ak intervened. The giant Awgwas try to destroy Claus, and an entire chapter is devoted to the "great battle between good and evil" in which the Awgwa army (augmented by Asiatic Dragons, three-eyed giants of Tatary, Black Demons from Patalonia, and Goozzle-Goblins) is utterly destroyed.

We learn how Claus begins to make other toys and deliver them in widening circles. We are told how two deer advise him to make a sleigh and promise to draw it, and why he finds it necessary to go down chimneys. As his fame grows, people start calling him a saint, or Santa Claus. From the friendly Gnome King he obtains steel runners for a larger sleigh and bells to jingle as he travels.

As the years speed by, Claus grows old and fat. His hair and beard turn white; deep wrinkles form at the corners of his bright eyes. He is about to die when his foster mother Necile seeks the aid of Ak. Ak calls a council of the immortals. They vote unanimously to confer upon good Claus the Mantle of Immortality.

Although old Santa is restored to the vigor of youth, he keeps his aged appearance. The growing number of children around the world make it necessary for him to acquire four immortal assistants: Kilter, Peter, Nuter, and Wisk. When the newfangled chimneys get too small for Santa's stomach, his assistants begin carrying toys through walls. Parents and toy shops are enlisted as aids to meet the demands of an expanding world population. "The more the merrier!" cries Santa.

On three later occasions Baum wrote again of Santa Claus. In the December 18, 1904, episode of a newspaper series called "Queer Visitors from the Marvelous Land of Oz," the Scarecrow, the Tin Woodman, and the Wogglebug make toy replicas of themselves and carry them to the Laughing Valley to give to Santa. The Delineator, December 1904, printed Baum's short story "A Kidnapped Santa Claus." It tells how five Daemons kidnap Santa, how he escapes, and how an army assembled by his four assistants repulses the Daemons.

In The Road to Oz, Santa Claus is the most distinguished guest at Ozma's birthday party. "Hello, Dorothy; still having Adventures?" the saint asks Dorothy Gale when they first meet. "And here's Button-Bright, I declare! What a long way from home you are; dear me!"

When Dorothy expresses surprise that Santa knows Button-Bright's father, the saint gives the Wizard a sly wink and says, "Who else do you suppose brings him his Christmas neckties and stockings?" (Santa's wink is one of those marvelous little touches that Baum liked to put in his books to amuse older readers.)

After chatting with Polychrome, the Scarecrow, the Tin Woodman, and Shaggy, Santa mounts the Saw-Horse for a sightseeing ride around the Emerald City. At the royal banquet it is Santa who makes the principal speech and leads everyone in a toast to Ozma. It is Santa, riding the Saw-Horse, who later leads the guests in a great parade through the streets of the Emerald City. Our last glimpse of him is when the Wizard blows one of his mammoth soap bubbles around the saint to float him home.

Oz and Santa Claus! The Ozzy old fellow was certainly at home there, and one assumes that he makes frequent visits to the Emerald City to see Ozma and Dorothy and their friends. Most Americans, who have not yet heard of the Laughing Valley or the Forest of Burzee, still suppose that the jolly saint lives at the North Pole. It is a quaint superstition that does little harm, but we Ozmapolitans know better.

17. *TALES OF THE LONG BOW*

G. K. Chesterton's *Tales of the Long Bow*, one of his long-forgotten books, is in my opinion among the best of his collections of short stories. My favorite tale is about the unthinkable theory of Professor Green. It tells of a great astronomical discovery by the professor that is central to a plot only G.K. could have thought of.

What follows is a chapter on *Tales of the Long Bow* in *The Fantastic Fiction of Gilbert Chesterton*, published in 2008 by Canada's Battered Silicon Dispatch Box.

G.K., as all Chestertonians know, had a great sense of humor. It comes through in almost all his writings, even when he is serious about philosophy and religion. There is that wonderful ending of *Orthodoxy* in which he fancies, with reverence, that when Jesus walked the earth he concealed his mirth.

Tales of the Long Bow is Chesterton at his funniest. In the tradition of Baron Münchausen, and Lord Dunsany's stories about Jorkens, its tales concern what G.K. calls things impossible to believe, "and, as the weary reader may well cry out, impossible to read about." Carefully constructed, written in a style more informal than usual, the tales are a delight to read and the book hard to put down.

The first story begins with a mystery. Why is the conventional Colonel Crane, an army veteran of the First World War, wearing to church a hat that is a hollowed-out green cabbage? Why did he smash his top hat, put it on the scarecrow in his garden, and for almost a week wear the cabbage hat everywhere? All who see him are astonished but too polite to ask why he seems to have gone mad.

It turns out that the Colonel had vowed to his lawyer friend Robert Owen Hood (Robert Owen was a famous Welsh socialist) that if Hood managed to do a certain seemingly impossible feat, he would eat his hat. When Hood finally did what Crane believed couldn't be done, he fulfilled his rash promise by turning a large boiled cabbage into a hat, and finally eating it at a ceremonial dinner.

As he so often does, Chesterton slips a romance into his narrative. Audrey Smith is an attractive art student who praises Crane for having the courage to appear a fool just to avoid going back on a rash vow. The two fall in love, and we learn in chapter 3 that they soon married.

Some lines reminiscent of Thorstein Veblen enter the story when Audrey extols the intrinsic beauty of a cabbage. "You want to apologize," she says, "for not wearing that stupid stovepipe covered with blacking, when you went about wearing a coloured crown like a king."

> "Literary people let words get between them and things. We do at least look at the things and not the names of the things. You think a cabbage is comic because the name sounds comic and even vulgar; something between 'cab' and 'garbage,' I suppose. But a cabbage isn't really comic or vulgar. You wouldn't think so if you simply had to paint it. Haven't you seen Dutch and Flemish galleries, and don't you know what great men painted cabbages? What they saw was certain lines and colours; very wonderful lines and colours."

I found in my copy of the book a yellowed clipping from *The New York Times* of November 12, 1959. It tells how James Mitchell, then Secretary of Labor, ate his "hat" after making a rash vow that he would do so if October's unemployment figures were not what he predicted in April. It would be the hat his critics accused him of talking through. When his prediction failed he had a white layer cake prepared, in the shape of a fedora, with mocha icing and a wide hat band of chocolate.

There is a photograph of Mitchell on the steps of the Labor Department Building, stuffing a piece of cake in his mouth. The imitation hat is on a table beside him.

Exactly what was the impossible thing that Owen Hood managed to do? That is the topic of G.K.'s next chapter.

Colonel Crane made his rash vow when his lawyer friend Hood was fishing on a small island in an upper portion of the Thames. "You may know a lot," Crane says, "but I don't think you'll ever set the Thames on fire. I'll eat my hat if you do."

How and why Hood did indeed set the Thames on fire is the burden of this tall tale. It involves an enormous brick factory for making a new kind of hair dye. It has been built on the banks of the Thames near where Hood liked to fish. The factory's owner is a wealthy captain of industry named Sir Samuel Bliss. His factory is polluting the air with smelly smoke, and poisoning the Thames with its disposal of excess oil and grease.

Associated with Bliss is Professor Hake (fake?). He has published research on how a thin layer of oil on a river acts as a "protective screen" that keeps the water in a purer condition. He is also the author of a book claiming that cosmetics are strongly hygienic, greatly improving the health of one's facial skin.

There is again, as so frequently in Chesterton's fiction, a romantic subplot. While Hood as a boy is fishing on the Thames islet, a girl named Elizabeth Seymour approaches with a batch of bluebells in her hand. When she accidently drops them into the river, Hood gallantly jumps in the water to rescue them. Instantly he is in love. Not until six years later do they meet again near the site of the abominable factory.

Hood professes to be an admirer of Horace Hunter, a friend of Bliss, who is running for Parliament. Actually he despises Hunter and his conservative opinions. On the eve of the election he organizes a torchlight parade. However, instead of leading the parade, he flings his torch into the oily Thames, where it sets the river on fire.

In a passionate but ironic speech to the crowd, before he throws the torch, Hood repeats three times the line from Edmund Spenser's "Flow gently, sweet Thames, until I end my song." Decades later Thomas Wolfe, in a great hymn to America's rivers (in his novel *Of Time and the River*), does exactly the same thing. He intersperses his long litany of rivers with the same line. Did Wolfe pick up this refrain from G.K., I wonder?

In my copy of *Tales of the Long Bow* I found a short clipping, the source of which I failed to record. It quotes the following paragraph from *The Illustrated London News* of March 3, 1906:

> Somebody really did set the Thames on fire. The floods of some oils loosened on the elder flood caught fire and gave an impression as if the water had been turned into flame.

Perhaps it was this event that gave G.K. the idea for his wild tale.

Chesterton and his friends were leaders of a political movement called distributism. In essence it recommended a more even distribution of wealth that would lessen the gap between the very rich and very poor. It was not socialist, but neither did it defend unrestrained capitalism. Aspects of distributism pervade several tales of the long bow—chapter 3 in particular.

Politicians in a certain county of England have, for unclear reasons, declared a ban on the raising of pigs. The ban is especially hard on John Hardy, innkeeper of the Blue Boar, who keeps on hand a supply of pigs to provide bacon for his hearty breakfasts of bacon and eggs.

Captain Hilary Pierce, a handsome young aviator, is in love with Joan, the innkeeper's daughter. Angered by the new laws against pigs, he decides to publicize opposition to the ban by violating it in every way he can think of. He owns a pet pig. He does time in jail for driving pigs into the county. He takes pigs on trains, disguised as wild animals or as human invalids on their way to being "cured." His ultimate act of folly is to create a huge Zeppelin in the shape of a pig. As a gift to Joan, the Zeppelin flies over the country and from it dozens of pigs descend in parachutes! It is a stupendous news event that violates the old proverb that says pigs can't fly!

The tall tale comes to an abrupt but happy ending. The ban against pigs is mysteriously lifted. It turns out that the laws were the work of a mild-mannered American capitalist from Michigan named Enoch Oates. He had connived with British authorities to ban pig raising because it would destroy his competition. Oates was "the biggest pork-packer and importer in the world." When the ban failed to hold, thanks to Captain Pierce's shenanigans, Oates changed his mind, ordered the ban lifted, and went into a different business.

Joan and her father think Pierce mad, but Joan loves him anyway. Not until chapter 5 do we learn that they were married, and learn also the nature of Oates's new business. Before then we are told the tall tale of Parson White's companion.

The Reverend Wilding White is an eccentric vicar of an Anglican church in the country town of Ponder's End. He is a bachelor who enjoys practical jokes and writing cryptic letters that begin "Yours truly" and end with "Dear sir." Colonel Crane, Owen Hood, and Captain Pierce are puzzled by a letter White has sent to Hood. It repeatedly speaks of his companion Snowdrop without revealing who Snowdrop is.

At one point White refers to people being alarmed if Snowdrop ever "took it into her head to walk about on two feet." This suggests to the three friends that Snowdrop might be a baby or a four-legged animal, perhaps a dog or cat, or maybe a pony or donkey. Pierce speculates that White may have converted to Spiritualism and Snowdrop could be a departed soul.

Intrigued by the mystery, Pierce decides to fly his plane to Ponder's End to find out just who Snowdrop is. His trip is a disaster. He returns more mystified than ever.

I'll not spoil the reader's pleasure giving away Snowdrop's identity, but I will provide a hint. The whimsical White has promised to bring something to sell at the village's White Elephant sale. The sale is being run by a tall, imposing woman whom Parson White at one time hoped to marry.

Snowdrop, by the way, is the name of Alice's kitten in Lewis Carroll's first *Alice* Book. The name also appears in Chesterton's great fantasy *The Man Who Was Thursday*. A mysterious character called Sunday, while being chased by the police through London, drops before his pursuers a nonsense note signed: Snowdrop.

In chapter 5 we are given a glimpse of the four friends—Captain Crane, Owen Hood, Hilary Pierce, and the Reverend Wilding White—when they were young. United by their propensity for crazy behavior, they had formed a social club called "The Lunatic Asylum." After the previous chapter's events they decided that what their little club most needed, as a kind of foil, is a sane man who will be shocked by their lunacies.

They want a man who is conventional and stupid, a solid, wealthy businessman. "In a word," says Hood, "I want a fool." They all agree that Enoch Oates, who made a fortune buying and selling pigs, is just the sort of person they need. "Oates is not a man I hate," says Hood. "He's a simple sincere sort of fellow . . . a thief and a robber of course, but he doesn't know it."

They invite Oates to dinner and he accepts. A nonstop talker, he delivers a monologue telling how he purchased hundreds of thousands of pigs, and from their ears obtained artificial silk to make attractive purses for women. Called Pigs' Whisper Purses because they came from ears, they were a huge success.

"And we thought we were dotty!" Pierce later exclaims in a burst of good distributist rhetoric. "American business rises to a raving idiocy compared to which we are as sane as the beasts of the field."

Colonel Crane, back from travels around the world, reminds his friends that in all cultures, including savage tribes, the sane man is the one who follows tribal customs. The best man is the man who wears a ring in his nose if nose rings are commonplace in his tribe. Oates is a good tribesman. He wears the nose rings of American business. "Nose rings are funny to people who don't wear 'em," Crane continues. "Nations are funny to people who don't belong to 'em." Oates is a perfectly normal, well-adjusted, admirable American. He is a good husband with a loving wife, a good father, and totally oblivious to the madness of capitalism.

Hilary Pierce visits Oates at his London hotel. He calls Oates's attention to the fact that when he was fighting competition by persuading British politicians to ban pig raising, he did great damage to his (Pierce's) wife and her innkeeper father who depended on pigs for his livelihood. "Can you really justify the way in which your romance nearly ruined their romance?"

"That's a mighty big question," Oates replies, "and will take a lot of discussing." The discussing occurs in the story's next chapter.

Of all the short stories by Chesterton, the one about Professor Green's unthinkable theory is my all-time favorite. The chapter opens with a sequel to the previous one. We learn that in a single session with Oates, Captain Pierce somehow convinced him to abandon his greedy capitalism and become a convert to what G.K. called distributism!

At the center of any free society, distributionists believe, everyone should own property. It seems that Oates had never before thought about private property. After seeing the light, he devotes his energy and his millions to providing farmers throughout the country with their own farms! Pierce's friend Owen Hood—we learn later—actually acquired a farm, which he surrounded with a moat and drawbridge bearing the motto "The Englishman's House Is His Castle."

Pierce and Crane, on a walk through the countryside, pass the home of Professor Oliver Green, a distinguished astronomer. Green is scheduled to deliver a paper at a congress of astronomers—a talk in which he will announce his new theory.

The new theory is equivalent to Einstein's general theory of relativity. Because all motion is relative to a fixed frame of reference, it doesn't matter what is chosen for the fixed frame. As Green explains to Pierce,

"When you run out of a village in a motor car," it would be equally true to say "the village ran away from you."

Later Green explains his theory to Margery Dale, the daughter of a farmer who lives near Green. There is no reason why one can't choose the earth as the fixed point and say the sun goes around it. Although I suspect Chesterton did not realize it, this is precisely the heart of Einstein's general theory. The equations of relativity are exactly the same if you assume that the sun goes around the earth. "I always *thought* it looked like that," Margery says. In the same way one can say the earth goes round the moon. It is only on grounds of simplicity that we choose the earth as the fixed frame. "Splendid!" Margery exclaims. "So the cow really does jump over the moon!"

Green starts to quote the next line of the nursery rhyme, "The little dog laughed . . ." when he suddenly stops talking and startles the girl with a burst of laughter. It was as surprising as if a nearby cow had laughed. Professor Green has experienced an epiphany. "All is beautiful," he says. "You are beautiful."

Green had been scheduled to speak on "Relativity in Relation to Planetary Motion." He has changed his mind. He sends to officials of the congress a telegram saying that instead of his promised topic he will announce his recent discovery of an entirely new planet!

Colleagues eagerly await this revelation. As his lecture proceeds they become more and more bewildered. Green starts by describing the exotic fauna and flora of the new planet, including details impossible to see with any telescope. He speaks of "towering objects which constantly doubled or divided themselves until they ended in flat filaments, or tongues of a bright green color." He describes monsters "resting on four trunks or columns which swung in rotation, and terminated in some curious curved appendages."

Captain Pierce, sitting in a front row, yells out, "Why, that's a cow!"

"Of course it's a cow," Green replies. Then he began waving his arms, calling his listeners "a pack of noodles who had never looked at the world they were walking on." This was followed by an outburst in praise of female beauty.

At that moment the meeting's chairman decides that Green has gone mad. There are efforts to remove him from the platform. He is rescued by Captain Pierce, who shouts that Green is the only sane man in the room, and takes him away in his airplane.

I despair of trying to summarize the book's last two chapters. The Earl of Eden has become Prime Minister. (No relation to Sir Anthony Eden, who would lead England many decades later.) When Enoch Oates began giving farms to farmers it generated a flood of discontent among rustics throughout England. In a drastic effort to combat this trend, Lord Eden decides to nationalize all the nation's land. Wealthy landowners will have their estates confiscated, then handed back to them to manage.

This nationalization of land triggers a revolution led by none other than Colonel Crane and his friends. They now call themselves the League of the Long Bow. Bellew Blair is their scientific genius. It was he who constructed the pig zeppelin, and who assisted Pierce in rescuing Oliver Green.

The war that ensues is even more preposterous than the battles in *Napoleon of Notting Hill* and *The Flying Inn*. Blair designs a gigantic zeppelin in the shape of a castle. From it he showers letters and propaganda leaflets all over the land. The huge "castle in the air" is kept in a deep underground laboratory where Blair fabricates other gas bags as well as strange weapons of warfare.

Throughout England peasants eagerly join the rebellion. Led by the Long Bowers, they quickly overthrow an incompetent, bumbling government to establish a new England based on the ideals of distributism! There are hints that the astronomer Professor Green, who has joined the Long Bowers, is enjoying a romantic relation with Margery, the farmer's daughter.

To convey something of the madness of this agrarian revolution, there is a scene in which soldiers of the Long Bow find a way to bend enormous trees to make catapults that pelt the enemy with darts and stones. There is even a suggestion that the evil scientist Hake (of chapter 2) has invented a powerful bomb. He offers the government "a new explosive capable of shattering the whole geological formation of Europe and sinking those islands in the Atlantic." Fortunately he is "unable to induce the cabman or any of the clerks to assist him in lifting it out of the cab."

Tales of the Long Bow was published in England by Cassell in 1925, and in America the same year by Dodd, Mead. All eight chapters were first serialized in *Storyteller* from June 1924 through March 1925. The book is included in volume 8 of the Ignatius Press edition of Chesterton's *Collected Works*. Donald Barr, the book's editor, provides an introduction and a wealth of enlightening footnotes.

A letter from John Peterson informed me that in 1908 Chesterton contributed to *The Daily News* a short fable titled "The Long Bow." It is about a group of four men who form a "League of the Long Bow" to pay tribute to England's "double bond" of its "heroic archery" combined with the "extraordinary credulity of its people." The fable is reprinted in G.K.'s *Alarms and Discursions* (1910), and in volume 14 of Chesterton's *Collected Works* (San Francisco: Ignatius Press, 1990).

18. WHEN YOU WERE A TADPOLE AND I WAS A FISH

There are endless examples of what the critic Burton Stevenson called "one-poem poets." These are poets who published many poems, only one of which became famous. Examples include the authors of "Casey at the Bat," "The Night Before Christmas," "The House by the Side of the Road," "Out Where the West Begins," "The Lost Chord," "The Old Oaken Bucket," and scores of other popular poems that will probably outlast the entire output of Ezra Pound.

Langdon Smith, author of a poem called "Evolution," was an extreme case. As far as anyone knows, it was the *only* poem he ever published!

Incredibly, I seem to be the only person who has ever deemed it worthwhile to check on who Langdon Smith was. My article about him, titled by the first line of his ballad, appeared in *The Antioch Review* (Fall 1962). It is reprinted here with a postscript updating what I have since learned about Smith and his poem. I offer $300 to anyone who can send me a copy of a publication of "Bessie McCall of Suicide Hall."

In 1941, T. S. Eliot startled the literary world by editing an anthology called *A Choice of Kipling's Verse*. Graduate students of English literature, who had been ashamed to admit that they could recite whole chunks from "Mandalay," "Boots," "Gunga Din," and "Danny Deever," suddenly found the courage to transfer their copies of Kipling from that dark, inaccessible corner of the bookcase to a shelf where they could be seen by visitors.

In the introduction to his anthology, Eliot proposed a useful distinction between "poetry" and "verse." Verse, he said, has a simple, metrical beat; it expresses clear, unambiguous ideas. Content and form can be grasped completely, or almost completely, on first reading. Poetry differs in degree. Its sound patterns and ideas are subtler, less lucid, impossible to understand fully on first reading. The poem's richness grows with rereadings.

Eliot did not intend the distinction to be invidious. Poetry and verse are two different things, they express different intentions. Kipling did not *try* to write poetry. What he wanted to do he did, and did magnificently. When he wrote (in "Danny Deever"), "What's that that whimpers over'ead?" the word "whimpers," said Eliot, is "exactly right." (Perhaps Eliot had Kipling's use of this word in mind when he used it himself, at the close of *The Hollow Men*, to describe the way the world ends.)

In short, Eliot thought Kipling wrote great verse.

In my opinion, Langdon Smith's "Evolution" is great verse. Not as good as Kipling's, perhaps, but good enough to merit critical attention. As to its popularity, there is not the slightest doubt. It has been reprinted in dozens of anthologies of "best loved" poems, usually—as is often the fate of poems of this sort—with badly garbled lines. (In Hazel Felleman's *Best Loved Poems of the American People*, for example, there are at least six flagrant errors.) It would be hard to find a geologist or biologist who has never read the poem; moreover, all sorts of unlikely people fall under its spell. Many years ago I sat at a table, in a Thompson's restaurant in Chicago, with a group of magicians attending a magic convention in the city. It was 3 a.m. Somehow the word *evolution* was mentioned. Harry Blackstone, one of the world's great stage conjurors, put down the deck of cards in his hand, cleared his throat, and recited Langdon Smith's poem from beginning to end. There was no special reason he had memorized it. The poem had simply caught his fancy when he was a young man.

Literary critics are seldom interested in "verse" unless the author is a famous writer (e.g., Edgar Allan Poe), so it is not surprising that they have nothing at all to say about Langdon Smith. No one, so far as I can discover, has ever written an article about him. In fact, the most remarkable thing one can say about Smith is that almost nothing has been said about him.

He was a newspaper reporter for the Hearst papers in New York City. There are two meager sources of biographical information: a short entry in Who's Who in America, 1906–07, and an obituary in the New York American, April 9, 1908, page 6.

The Who's Who sketch is tantalizingly vague. Smith was born in Kentucky (no town is mentioned) on January 4, 1858, had a common school education at Louisville (1864–72), and was married on February 12, 1894, to Marie Antoinette Wright. (His obituary speaks of her as a Louisville girl.) During his boyhood he served in the Comanche and Apache wars; later he was a correspondent in the Sioux war. He went to Cuba for the New York Herald when war broke out in 1895. When the United States declared war on Spain, he was a correspondent in Cuba for the New York Journal. He was (the sketch continues) at the bombardment of Santiago, on the hill with the Marines at Guantánamo, and present at the battles of El Caney and San Juan. The sketch closes: "Author: On the Panhandle; also short stories. Address: 154 Nassau Street, New York."

No book called On the Panhandle was ever published. Was it an unprinted manuscript? Novel or nonfiction? Is the title a hoax? Were his short stories ever published? I have been unable to find the answer to any of these questions.

Smith's obituary in the New York American adds little except a photograph and some details about his illness and death. The picture is that of a handsome, dark-mustached man. He is called "one of the best known of newspaper writers in the country." For ten years he had been on the newspaper's staff. His last signed articles of importance had been on the second trial of Harry K. Thaw and the departure of the U.S. naval fleet on its first round-the-world cruise. His interview with the then secretary of war, William Howard Taft, had appeared on March 23, 1908.

Biographical details in this obituary are obviously copied from *Who's Who*, though the writer has embroidered some of them. Thus: "His best known work was entitled *On the Panhandle*, but he had also written many short stories which had a considerable vogue." In addition to his poem "Evolution," Smith had also written (so the obituary reads) a "familiar" poem entitled "Bessie McCall." (This poem is not listed in *Granger's Index to Poetry*; I have been unable to discover when or where it was printed.) Friends called him Denver Smith, a nickname alluding to his earlier days as a telegraph operator in Denver. His death on April 8 occurred at his home, 148 Midwood Street, Brooklyn. He was fifty. There is no mention of children or other survivors.

On the following day the *American* reprinted the entire poem "Evolution" on its editorial page, followed by a tribute to Smith. "He had no mannerisms, no affectations, no sentimentalities. Keenly enjoying life, whether of the plains or the city, he invested it with enjoyment for those who knew him and for those who read what he wrote." Of his poem, the tribute says: "Mr. Smith wrote and rewrote it many times before he was satisfied with it."

A check through the clippings on Smith that are on file in the reference room of the *New York Journal-American* turned up little of interest. Two thin folders on him contained only his obituary, a sampling of a sports column that he wrote, and a column about him, by Hype Igoe, dated April 21, 1939. Mr. Igoe, who had known Smith, recalls a purple passage from Smith's celebrated account of the testimony of Evelyn Thaw on the witness stand to defend her husband; he writes about Smith's friendship with Jim Corbett in the days before Corbett defeated John L. Sullivan for the world's heavyweight boxing crown. Apparently Smith was at one time sports editor of the paper. Igoe mentions Smith's other poem, giving it a fuller title and one that suggests it has been deservedly unremembered: "Bessie McCall of Suicide Hall."

The first appearance in print of "Evolution" as a complete poem was on a page of classified advertisements in the morning Hearst paper, then called the *New York Journal and Advertiser*. The exact date is not known. A paragraph heading a reprint of the poem (in a magazine called *The Scrap Book*, April 1906, pp. 257–59) reads:

> History records that in 1895 Mr. Langdon Smith, at that time connected with the Sunday edition of the New York *Herald*, wrote the first few stanzas of the following poem. They were printed in the *Herald*. Four years later, having joined the staff of the New York *Journal* in the interim, Mr. Smith came across the verses among his papers, and, reading them over, was struck with a sense of their incompleteness.
>
> He added a stanza or two and laid the pen aside. Later he wrote more stanzas, and finally completed it and sent it in to Mr. Arthur Brisbane, editor of the *Evening Journal*. Mr. Brisbane, being unable to use it, turned it over to Mr. C. E. Russell of the *Morning Journal*. It appeared in the *Morning Journal*—in the middle of a page of want "ads"! How it came to be buried thus some compositor may know. Perhaps a "make-up" man was inspired with a glimmer of editorial intelligence to "lighten up" the page.
>
> But even a deep border of "ads" could not smother the poem. Mr. Smith received letters of congratulations from all parts of the world, along with requests for copies. The poem has been in constant demand; and it has been almost unobtainable . . .

Two years later the poem was printed again in a quarterly magazine, *The Speaker* (September 1908, pp. 394–97). Since then it has been reprinted countless times in anthologies. So far as I know, the first anthology to include it was Edwin Markham's two-volume *Book of Poetry* (vol. 1, p. 352). "Strange to say," Markham comments, "this is the only poem of distinction that he is known (to me) to have written."

Strange also is the fact that Markham's version of "Evolution" supplies four lines missing from all other versions of the poem I have seen. The lines precede the poem's final four-line stanza and are as follows:

For we know that the clod, by the grace of God,
Will quicken with voice and breath;

And we know that Love, with gentle hand,
Will beckon from death to death,

It is unlikely that Markham would have added these inferior lines just to make the last stanza conform in length to the others. More likely they were in the first printed version. One wonders if Smith himself later blue-penciled them.

"Evolution" has been printed as a book or pamphlet at least four times. A hand-lettered booklet, with line drawings by an anonymous artist, bears neither date nor publisher's imprint. In 1909 a fifty-one-page hardcover version was issued by John W. Luce and Company, with an introduction by Lewis Allen Browne. (Browne is best known as the author of a popular book on religion, *This Believing World.*) Unsigned notes on the various geological terms mentioned in the poem are placed on pages opposite the stanzas. Browne's introduction gives the impression that he knew Smith, but obviously he did not. His "facts," from the sources already mentioned, add little of interest. In 1911 another hardcover edition was published by W. A. Wilde and Company, with decorative page borders by Fred S. Bertsch. This edition apparently enjoyed wide distribution, because copies of it, until a few years ago, were common in secondhand bookstores. A Danish translation by Jens Christian Bay (no illustrations) was privately printed in Holstebro, Denmark, in 1930.

How can one explain the poem's continuing popular appeal? First of all, there is the strong singsong rhythm that makes the poem so easy to memorize, so effective to recite. Its stanzas are constructed skillfully, and there are passages (as Eliot noted of Kipling's verse) that reach the intensity of poetry. Second, the poem conveys what Darwin had in mind when he wrote, at the close of his *Origin of Species,* "There is grandeur in this view of life." The epic surge of evolution, from its humble beginning in the dark sea to the mellow light of Delmonico's, is caught in this poem as it has not been caught in any other poem before or since.

Finally, Smith has given evolution a strong religious cast. Did he himself believe in transmigration? It would be interesting to know.

The theme of lovers reuniting in successive reincarnations, throughout the earth's long geological history, was a common one in popular fantasy novels of the late nineteenth century. Smith handles the theme so playfully, however, that it is hard to tell how serious he is. It is precisely his ambiguous touch here that makes it possible for the poem to be enjoyed by all readers, regardless of their beliefs about the soul's existence before birth or after death. Lewis Browne, for example, was a thoroughgoing naturalist, yet he thought the poem had "the ring of genius," and that its "crowning glory" was the way in which Smith "interwove throughout his masterpiece of imagination the golden thread of romance."

Browne closes his introduction by noting that Smith's wife died within five weeks after her husband's death, a fact that his friends took to indicate the strong bond between them. I have not been able to verify this. It may be truth, it may be a myth inspired by the poem. It is one of the many curious mysteries that continue to surround the poem and its author.

There is, at least for me, something haunting and sad about the lives of "one-poem men," as Burton Stevenson calls them in his book on *Famous Single Poems*, individuals who achieve immortality by writing one great piece of popular verse and nothing else. Because of the "nothing else" they almost literally vanish from history. "Casey at the Bat" was first published in the *San Francisco Examiner* (another Hearst paper, by the way) in 1888. The author, Ernest Lawrence Thayer, was, like Smith, a journalist. It is this country's best-known humorous poem. What sort of man was Thayer? What else did he write? "A Visit from St. Nicholas," another immortal poem, appeared anonymously in 1823 in the *Sentinel*, a newspaper in Troy, New York. The author was Clement Clarke Moore, professor of Oriental languages at the General Theological Seminary of New York. He is remembered today only because he scribbled down these lines to read to his children on Christmas Eve. He considered them so trivial that for twenty years he refused to admit he had written them.

Robert Graves, writing about "one-poem men" in his book *On English Poetry*, divides them into two types: the born poets, so frustrated by environment that only once in their lives are they able to break through and create a poem, and those who are not poets at all, but who "write to express a sudden intolerable clamour in their brain." Having once expressed themselves, they have no further need of poetry. The first type is not hard to understand. The second type, if such there be, is not so easy to understand. I should think that there is a field here for investigation by those psychologists concerned with creativity.

Scholars have done their work on Moore and Thayer (both are included in Burton Stevenson's book); Smith is still wrapped in undeserved anonymity. Even his name is appropriately anonymous: Smith. But smiths are artisans, and Langdon Smith was a credit to his clan. He fashioned for himself what must have been a colorful life, and on at least one occasion he hammered out a colorful ballad. It will probably be chanted long after the efforts of many contemporary poets, with enormous reputations, are forgotten by everybody except the historians of literature.

EVOLUTION

LANGDON SMITH

I

When you were a tadpole and I was a fish,
In the Paleozoic time,
And side by side on the ebbing tide
We sprawled through the ooze and slime,
Or skittered with many a caudal flip
Through the depths of the Cambrian fen,
My heart was rife with the joy of life,
For I loved you even then.

II

Mindless we lived and mindless we loved,
And mindless at last we died;
And deep in a rift of the Caradoc drift
We slumbered side by side.

The world turned on in the lathe of time,
The hot lands heaved amain,
Till we caught our breath from the womb of death,
And crept into light again.

III
We were Amphibians, scaled and tailed,
And drab as a dead man's hand;
We coiled at east 'neath the dripping trees,
Or trailed through the mud and sand,
Croaking and blind, with our three-clawed feet
Writing a language dumb,
With never a spark in the empty dark
To hint at a life to come.

IV
Yet happy we lived, and happy we loved,
And happy we died once more;
Our forms were rolled in the clinging mold
Of a Neocomian shore.
The eons came, and the eons fled,
And the sleep that wrapped us fast
Was riven away in a newer day,
And the night of death was past.

V
Then light and swift through the jungle trees
We swung in our airy flights,
Or breathed in the balms of the fronded palms,
In the hush of the moonless nights.
And oh! what beautiful years were these,
When our hearts clung each to each;
When life was filled, and our senses thrilled
In the first faint dawn of speech.

VI

Thus life by life, and love by love,
We passed through the cycles strange,
And breath by breath, and death by death,
We followed the chain of change.
Till there came a time in the law of life
When over the nursing sod
The shadows broke, and the soul awoke
In a strange, dim dream of God.

VII

I was thewed like an Auroch bull,
And tusked like the great Cave Bear;
And you, my sweet, from head to feet,
Were gowned in your glorious hair.
Deep in the gloom of a fireless cave,
When the night fell o'er the plain,
And the moon hung red o'er the river bed,
We mumbled the bones of the slain.

VIII

I flaked a flint to a cutting edge,
And shaped it with brutish craft;
I broke a shank from the woodland dank,
And fitted it, head and haft.
Then I hid me close to the reedy tarn,
Where the Mammoth came to drink;—
Through brawn and bone I drave the stone,
And slew him upon the brink.

IX

Loud I howled through the moonlit wastes,
Loud answered our kith and kin;
From west and east to the crimson feast

The clan came trooping in.
O'er joint and gristle and padded hoof,
We fought, and clawed and tore,
And cheek by jowl, with many a growl,
We talked the marvel o'er.

X

I carved that fight on a reindeer bone,
With rude and hairy hand,
I pictured his fall on the cavern wall
That men might understand.
For we lived by blood, and the right of might,
Ere human laws were drawn;
And the Age of Sin did not begin
Till our brutal tusks were gone.

XI

And that was a million years ago,
In a time that no man knows;
Yet here tonight in the mellow light,
We sit at Delmonico's;
Your eyes are deep as the Devon springs,
Your hair is as dark as jet,
Your years are few, your life is new,
Your soul untried, and yet—

XII

Our trail is on the Kimmeridge clay,
And the scarp of the Purbeck flags,
We have left our bones in the Bagshot stones,
And deep in the Coraline crags;
Our love is old, our lives are old,
And death shall come amain;
Should it come today, what man may say
We shall not live again?

XIII

God wrought our souls from the Tremadoc beds
And furnished them wings to fly;
He sowed our spawn in the world's dim dawn,
And I know that it shall not die;
Though cities have sprung above the graves
Where the crook-boned men made war,
And the ox-wain creaks o'er the buried caves
Where the mummied mammoths are.

XIV

Then as we linger at luncheon here,
. *O'er many a dainty dish,*
Let us drink anew to the time when you
Were a Tadpole and I was a Fish.

POSTSCRIPT

As far as I know, my tribute to Smith continues to be the only article ever written about him or his poem. I had intended to do a series of essays on "one-poem poets," with an eventual book in mind, but I wrote only three more. My paean to Ernest Thayer and "Casey at the Bat" first appeared in *Sports Illustrated*, then later was expanded to make the introduction to my book *The Annotated Casey at the Bat* (1967). The story of Clement Moore's Christmas poem will be found in the introduction I wrote for the Dover reprint of L. Frank Baum's fantasy *The Life and Adventures of Santa Claus* (see chapter 16 of this collection) and in the introduction to my *Annotated Night Before Christmas* (Amherst, NY: Prometheus, 2003).

At the time Smith wrote his ballad, the theme of couples having been lovers in past incarnations was common in both poetry and fiction. A notable verse example is Kipling's "The Sack of the Gods." It has exactly the same stanza form as Smith's ballad, and contains such lines as "I was Lord of the Inca race, and she was Queen of the Sea." W. E. Henley's poem "To W. A.," which closes with the lines

"When I was a king in Babylon, and you were a virgin slave," was also widely quoted in Smith's day. A listing of novels with reincarnation love themes, which were bestsellers in the early 1890s, would run to dozens of titles. Smith's originality was in combining reincarnation with evolution. If there are earlier examples of this in English fiction or poetry, I have not come across them. Nor do I know of any memorable examples that came later. There must have been several comic parodies of Smith's ballad, but I know of only one, a mediocre baseball parody included in Grantland Rice's *Only the Brave* (1941).

I was tempted to annotate the geological terms in Smith's poem, but decided this would add nothing to the poem. Delmonico's, in stanza 11, is of course the famous Manhattan restaurant where fashionable people once dined. In Smith's day it was at the corner of Fifth Avenue and Forty-fourth Street, where it remained until it expired in 1923. An entire book about it, *Delmonico's: A Century of Splendor*, by Lately Thomas, was published in the late sixties.

I am indebted to Janet Jurist for tracking down the first appearance in print of "Evolution." It was in the *New York Herald*, September 22, 1895, where it was divided into stanzas of four lines each, and illustrated with five drawings. There was no byline! Several years later Smith added seven new quatrains to create the ballad's final version in the *New York Morning Journal* on a date still unknown. If you break the stanzas given here into quatrains and number them 1 through 27, the stanzas not included in the first publication are numbers 15 through 19, 25, and 26.

If you number the eight-line stanzas from 1 through 13, the following words were changed from the original version:

Stanza 1, line 3, *sluggish* was changed to *ebbing*.
4:6, *wrapped* to *bound*.
5:4, *moonlit* to *moonless*.
10:8, *brutish* to *brutal*.
12:8, *meet* to *live*.

All five changes are obvious improvements.

Google has a lengthy section of letters from people telling of their great love for this poem and about relatives who could recite it from beginning to end. Several mention Jean Shepherd's recital of "Evolution" on his popular radio program.

Smith's widow, grief-stricken over her husband's death on April 8, 1908, committed suicide on June 11 by taking poison. It was her second attempt. She died at her home in Flatbush, New York. See *The New York Times*, June 11, 1908, page 4.

PART VI

RELIGION

19. THE CURIOUS CASE OF FRANK TIPLER

It is hard to believe that Frank Jennings Tipler exists. He is a respected physicist at Tulane University and the author of many technical papers. He is also the author of two of the most outlandish books ever written about religion: *The Physics of Immortality* (1994) and *The Physics of Christianity* (2007).

A recent Catholic convert, Tipler is convinced that all the great miracles of the New Testament can be explained by the laws of physics! Miracles are not, as Christians have always believed, cases of God's *violating* natural laws, but cases of God's *using* natural laws. My review of Tipler's 2007 book appeared in *The Skeptical Inquirer*'s March/April 2008 issue.

The *Physics of Christianity* by Frank Tipler, a mathematical physicist at Tulane University, is a sequel to *The Physics of Immortality*, a bestseller in Germany before it was published here in 1994 by Doubleday. In that book, Tipler argued that anyone who understands modern physics will be compelled to believe that at a far-off future date, which Tipler calls the Omega Point (borrowing the term from the Jesuit paleontologist Teilhard de Chardin), God will resurrect every person who lived, as well as every person who could have lived! Our brains will be preserved as computer simulations and given new spiritual bodies to live happily forever in the paradise described in the New Testament.

In his new book, published in 2007 by Doubleday, Tipler goes far beyond his previous one. He claims that modern physics also provides reasonable explanations for the historical accuracy of all the central miracles of Christian faith, as well as the many alleged miracles that continue to take place, notably those associated with Catholic saints. "From the perspective of the latest physical theories," Tipler writes in his introduction, "Christianity is not a mere religion but an experimentally testable science." Roll over, Mary Baker Eddy!

It is no surprise that Tipler has become a conservative, orthodox Catholic. On page 217 he attributes his conversion to the influence of the German Lutheran theologian Wolfhart Pannenberg.[1] "[He] spent fifteen years in a finally successful attempt to persuade an American physicist (me) that Christianity, undiluted Chalcedonian Christianity, might in fact be true and might even be proved to be true by science."

There are two ways, Tipler writes, to regard miracles:

1. They are, as David Hume famously maintained, alleged supernatural events that violate laws of science.
2. They are highly improbable events performed by God, but without violating any natural laws.

The second view is the heart of Tipler's new book.

One can think of Tipler as a Christian version of Immanuel Velikovsky. A devout orthodox Jew, Velikovsky explained the great miracles of the Old Testament by invoking the laws of physics (see "Creationism, Catastrophism, and Velikovsky," *The Skeptical Inquirer*, January/ February 2008). Thus, Joshua was able to make the sun and moon stand still in the sky because a giant comet erupted from Jupiter and passed close to earth, causing it momentarily to stop rotating. It also caused the Red Sea to part precisely at the moment Moses commanded it. The comet showered edible manna on Israel before it settled down to become Venus.

Velikovsky had no interest in New Testament miracles, unlike Tipler, who is concerned with New Testament miracles but is silent on Old Testament ones. It would be interesting to know what he thinks about the dreadful fate of Lot's wife or the agony of Jonah in the belly of a whale. Tipler has a natural explanation for every miracle of Christianity, including those not in the Bible but one infallibly validated by the Roman Church in 1950. All are caused by God, though "never ever" by abrogating any law of physics.

Tipler devotes chapter six to the Star of Bethlehem. The accuracy of Matthew's account is never questioned. The star was not a supernatural event, nor was it a conjunction of Jupiter and Saturn as some Bible commentators surmise. It was, Tipler assures us, a supernova bursting in the galaxy of Andromeda. God cleverly timed the nova so it would signal the birth in Bethlehem of his only begotten son.

Chapter seven reveals for the first time the dark secret of the Virgin Birth. It was a rare case of parthenogenesis! This is the technical term for births that lack male fertilization of a female egg. The phenomenon is fairly common among certain vertebrates, such as snakes, lizards, sharks, and turkeys; Tipler sees no reason why it can't occur in humans, and he suspects it actually does occur. He is convinced this happened with Mary. Moreover, he thinks Mary's parthenogenesis could be confirmed by careful analysis of Jesus's blood on the Shroud of Turin!

Tipler has no doubts about the genuineness of the Shroud. Two microphotographs of the blood are introduced, and Tipler claims that its DNA is consistent with Mary's virginity. True, the Holy Spirit played a mysterious role in the Virgin Birth, but the birth broke no biological laws. The Bible, Tipler reminds us, implies that Joseph did not believe his young wife when she denied that any man was involved in her being with child.

All conservative Christians believe Jesus was free of the original sin that resulted from the Fall, which has been passed on to all descendants of Adam and Eve. Catholics think that Mary, too, escaped original sin. (For Catholics it is heresy to reject the Immaculate Conception.) How does Tipler explain the way Jesus and Mary differ in this manner from all other humans?

Tipler's answer is wonderful. There must be genes that carry original sin! This could be verified someday, he writes, by first identifying the gene. Thus, failing to find evidence of the gene on the Shroud of Turin would explain the sinlessness of both Jesus and his mother.

(I am, dear reader, doing my best to keep a straight face while I summarize Tipler's convictions.)

Chapter seven is about Jesus's resurrection. Here Tipler plunges into technical regions of quantum mechanics (QM). He is a firm believer in what is called the "many worlds interpretation" of QM. All I need say here about this fantastic view is that it assumes the reality of a "multiverse" that contains an infinity of universes similar to our own. Millions of these parallel worlds contain exact duplicates of you and me. Tipler quotes Stephen Hawking as saying to him that the many worlds interpretation of QM is "trivially true."

If Hawking said this, I think he meant that the many worlds interpretation is a useful language for talking about QM, but its infinite parallel worlds are not "real" in the same way our universe is real. However, for Tipler they are very real. Denying the multiverse, he says "is the same as denying that 2+2=4" (Tipler, p. 16).

Here is a typical paragraph about Jesus's Resurrection:

I am proposing that the Son and Father Singularities guided the worlds of the multiverse to concentrate the energy of the particles constituting Jesus in our universe into the Jesus of our universe. In effect, Jesus' dead body, lying in the tomb, would have been enveloped in a sphaleron field. This field would have dematerialized Jesus' body into neutrinos and antineutrinos in a fraction of a second, after which the energy transferred to this world would have been transferred back to the other worlds from whence it came. Reversing this process (by having neutrinos and antineutrinos—almost certainly not the original neutrinos and antineutrinos dematerialized from Jesus' body—materialize into another body) would generate Jesus' Resurrection body.

Although Tipler has nothing to say about the resurrection of Lazarus and other revivals of the dead mentioned in the New Testament, presumably they have similar explanations.

Tipler also reveals, so help me, exactly how Jesus managed to walk on water. He performed this great magical feat by "directing a neutrino beam" downward from his feet. Similar neutrino beams account for his ascension into the clouds, as well as for how his resurrected body was able to dematerialize and rematerialize. Mary's assumption is similarly explained: Tipler recommends checking her tomb for tracks of nuclear particles that would have been generated by her assumption. Apparently, Tipler thinks her corpse floated into heaven from her tomb rather than from a funeral procession, as one legend has it.

Chapter nine describes how physics explains the Incarnation, and how it also can account for the real presence of the Lord's body and blood in the bread and wine of the Catholic Eucharist.

I will spare the reader accounts of Tipler's belief that within fifty years computers will surpass human intelligence, and how our organic brains will be replaced by computer emulations as the universe moves inexorably toward the Omega Point. When that point is reached, an evolving God will become omniscient in the sense of knowing everything that can be known and omnipotent in the sense of being able to do everything that can be done. As Thomas Aquinas taught, there are things God cannot do, such as create a world that contains logically impossible things like a triangle with four sides or a creature that is both a perfect human and a perfect horse. It is best, Aquinas adds, not to say there are things God can't do, but that there are things that can't be done.

Before fifty years have ended, Tipler warns us, Armageddon will be fought with weapons that will make nuclear bombs seem like "spitballs" (p. 254). There will be mass conversions of Jews to Christianity. Tipler dedicates his book "To God's Chosen People, the Jews, who for the first time in 2000 years are advancing Christianity." After Armageddon, Jesus will return in glory to reign over a new earth. How does Tipler know all this? Biblical prophecy says so! "Before the Second Coming," he writes (p. 369), "I would expect to see a Jewish Pope."

For a few moments, after finishing *The Physics of Christianity*, I began to wonder if the book could be a subtle, hilarious hoax. Sadly, it is not.

20. THE COMIC PRATFALL OF RICHARD ROBERTS

As a native of Tulsa, where I was born and raised, I have long been fascinated and amused by the antics of Oral Roberts and his handsome singing son, Richard. Like so many other Pentecostal Bible thumpers, Richard has become a victim of the love of conspicuous waste. In 2008 he resigned as president of Oral Roberts University, charged with misusing university funds to support his and his family's lavish lifestyle.

I tell the story of Richard's sudden downfall in the following chapter. It first appeared in *The Skeptical Inquirer* (March/April 2008).

> *Lay not up for yourselves treasures upon earth, where moth and rust doth corrupt and where thieves break through and steal.*
>
> —Jesus, Matthew 6:19

In November 2007, Oral Roberts's singing son, Richard, age fifty-nine, resigned as president of Tulsa's Oral Roberts University (ORU). For fourteen years, his presidency has been a Tulsa joke. Now three former university professors are suing Richard and ORU for allegedly illegally using university funds to maintain the lavish lifestyle of Richard and his second wife, Lindsay. Three somewhat similar lawsuits have since been filed by individuals.

On Thanksgiving Day, Richard said that although he was completely innocent of all charges, God told him to resign. As all Tulsans know, God often speaks directly to Richard and his father. On his television show, Richard routinely receives from the Lord what Pentecostals call the "word of knowledge." Richard will typically say, "There is someone watching this broadcast who is slowly losing her sight because of cataracts. God is healing your eyes at this very moment! Praise the Lord!"

In the early 1960s, God said to Oral, "Build me a university." After ORU was built, God spoke again: "Oral, build me a hospital." While the medical complex, which Oral called the City of Faith, was under construction, Oral had a vision of a nine-hundred-foot-tall Jesus. "He stared at me without saying a word. Oh, I will never forget those eyes. He reached down his hands under the City of Faith, lifted it, and said to me, 'See how easy it is for me to lift it!'" Posters went up near the site of the vision. "Begin the 900-foot Jesus Crossing," they warned.

What Tulsa didn't need was another hospital. Over the years, rooms at the City of Faith were mostly empty until Oral was finally forced to turn the hospital into an office building. Apparently God had given him bad advice.

Two enormous bronze praying hands stand on the ORU campus. One of the best of many Tulsa jokes about ORU tells of a time when a small earthquake toppled the hands. A construction crew was struggling vainly to get the hands upright when a man approached the foreman and said, "If you give me a quarter I'll show you how to get those hands back up." Amused, the foreman handed the man a quarter. He tossed it toward the hands. They instantly jumped up to grab it.

In 1987, Oral was back in the news. God told him he would call him home if he failed to obtain millions of dollars to pay off a mounting debt soon. Bumper stickers appeared all over town saying "Send Oral to heaven in '87." Sure enough, a whopping check from a Florida dog racetrack owner saved the day.

ORU is currently struggling to erase a debt of $50 million. In spite of this debt, Richard has allegedly been siphoning cash from ORU for years to support an outlandish lifestyle. University funds have supposedly been used for eleven costly remodelings of his mansion. Lindsay's closet is the size of a bedroom. In less than a year, she spent tens of thousands of dollars on clothes from a Tulsa store alone. She drives a white Lexus SUV and a red Mercedes convertible, both bought, maintained, and fueled, critics say, by ORU. A private jet owned by the university is used only by Richard, his family, and his friends. It is also believed that ORU paid $29,411 for the vacation of a daughter and her friends to Orlando, Florida, and to the Bahamas. It bought and maintains a stable of horses for Richard's three daughters. Phone bills for the family, all allegedly paid by ORU, run to $800 a month.

According to a document submitted by the three plaintiffs, "As of 2003 Richard Roberts was compensated at the following levels—$181,469 annual salary from ORU, in excess of $100,000 as vice president of City Plex, and an additional $41,530/year from the Oral Roberts Evangelistic Association." *The Oklahoman* (October 21, 2007) reported that in 2005 Lindsay, executive vice president of the association, was paid $77,012 by the association and $119,800 by its subsidiary Trico Advertising. In 2006 Oral Roberts received $83,505 as the association's trustee.

Richard's first wife, Patti, did not appreciate Richard's assurance, which he got from his father, that God had no objection to enormous wealth and conspicuous waste. Does not the 23rd Psalm say "I shall not want"? Yes, and the gospels tell how Jesus advised a rich young man to sell his possessions and give the money to the poor, and that Jesus said it was easier for a rope (in the consonantal Aramaic script, the words for *camel* and *rope* are spelled the same) to go through a needle's eye than for a rich man to enter heaven. After Patti divorced Richard, she wrote a book titled *Ashes to Gold* that draws a grim picture of Richard and their stormy marriage.

The lawsuit filed by the three professors alleges that twenty-nine photographs exist of Lindsay sitting in her sports car after midnight with a sixteen-year-old boy and smoking cigarettes. Several persons described the boy as Lindsay's "boyfriend." The lawsuit points out that Tulsa has a 10 p.m. curfew for all teenagers not in the company of their parents. I have no idea if all this is true, or if it is, what to make of it.

"Allegations against me in a lawsuit . . . are not true," Lindsay declared. "They sicken me to my soul."

Pentecostal evangelist Pat Robertson, a longtime friend of Oral, has made an undisclosed donation to help pay ORU's debt of $50 million. He has offered to pay more, provided ORU cleans up its act.

It's hard to imagine anyone less qualified than Richard to run a university. True, he has a PhD degree. And what college gave Dr. Roberts his doctorate? Yes, you guessed it—ORU. An overwhelming majority of the ORU faculty are jubilant to see Richard go. At last the Lord seems to have given him good advice.

21. WHY I AM NOT AN ATHEIST

So many best-selling books these days devoutly defend atheism, with Richard Dawkins's *The God Delusion* topping the list, that I was tempted to consider writing a book on philosophical theism. A philosophical theist is a person who believes in God but not in any organized religion. The list of such thinkers is long and distinguished. It starts with Plato, includes Kant, and in more recent times, William James, his friend Charles Peirce, the Spanish philosopher and novelist Miguel de Unamuno, and many others.

I decided against writing such a book on the grounds that I had already defended philosophical theism in two books. One is my confessional, *The Whys of a Philosophical Scrivener* (1983). The other is my crazy novel, *The Flight of Peter Fromm* (1973).

Both books are still in print. They surprised and dismayed many of my loyal readers. They had assumed that because I don't believe in astrology or homeopathy, or that Uri Geller can bend spoons by psychokinesis, and because I was one of the founders of *The Skeptical Inquirer*, I must be a secular humanist!

The following essay reprints chapter 13 of my *Whys*.

"If you listen to your heart," said the Philosopher, "you will learn every good thing, for the heart is the fountain of wisdom, tossing its thoughts up to the brain which gives them form."
—James Stephens, *The Crock of Gold*

Whenever I speak of religious faith it will mean a belief, unsupported by logic or science, in both God and an afterlife. Bertrand Russell once defined faith, in a broader way, as "a firm belief in something for which there is no evidence." If "evidence" means the kind of support provided by reason and science, there is no evidence for God and immortality, and Russell's definition seems to me concise and admirable.

Faith of this pure sort, uncontaminated by evidence, is easily caricatured. In "The Will to Believe" William James quotes a schoolboy remark: "Faith is when you believe something you know ain't true." No fideist accepts this, of course, but if we alter it to "Faith is believing something you don't know is true," it is not a bad definition.

In the Christian tradition, *faith* has two related but distinct meanings. One is that of nonrational belief, the sense adopted here; the other is that of trust. Trust in God presupposes faith in the belief sense. You can't trust a person unless you think that person exists. Throughout both Testaments of the Bible, faith almost always means trust. It has often been observed that nowhere does the Bible give arguments for God. You will look in vain for them in the preaching of Jesus. God's reality is taken for granted, never defended. I will not be concerned here with faith as trust, only with faith as belief.

The author of the Epistle to the Hebrews opens his famous chapter on faith, chapter 11, with a familiar definition that Edgar J. Goodspeed translates as follows: "Faith means the assurance of what we hope for; it is our conviction about things we cannot see." To a philosophical theist this is a superb definition of faith as belief, even though the chapter goes on to catalog instances of faith that are more examples of blind trust than belief, and which no non-Christian theist can accept as historical or even praiseworthy.

I do not, for example, believe that God ever drowned all men, women, and children on the earth (not to mention innocent animals) except Noah and his family. Even as a myth it is hard to admire the "faith" of a man capable of supposing God could be that vindictive and unforgiving. I do not believe that God asked Abraham to murder . his only legitimate son as a blood offering. I know how Abraham's obedience has been justified, and I have read Kierkegaard's little book about it, *Fear and Trembling*, but unlike Kierkegaard and the author of Hebrews I am under no obligation to find anything beautiful or profound in this abominable story. To those outside the Judeo-Christian tradition, Abraham appears not as a man of faith, but as a man of insane fanaticism. He would have done better to have supposed that he was listening to the voice of Satan.[1]

Jephthah, also mentioned in the eleventh chapter of Hebrews as a man of faith and uprightness, is even harder to admire. Since only orthodox Jews and Protestant fundamentalists now read the Old Testament thoroughly, let me urge you, if you are a liberal Christian, to look up Judges, also chapter 11, and see what you can make of this horror tale. Read how Jephthah made a rash vow that if he won a military victory over the Ammonites he would sacrifice whatsoever first came out of his house to greet him when he returned. That turned out to be his only child. The virgin girl so loved her father that she met him dancing and shaking tambourines. Read how the poor girl, upheld by her great "faith," cooperated with the demented judge and warrior. "O Jephthah, judge of Israel," exclaims Shakespeare's Hamlet, "what a treasure hadst thou!"[2]

The Old Testament God, and many who had great "faith" in him, are alike portrayed in the Bible as monsters of incredible cruelty. A philosophical theist, standing outside any religious tradition, can construct better models of God than Jehovah. Nevertheless, Hebrews 11:1, especially in the familiar phrasing of the King James Bible, remains a beautiful way of saying how faith, as a form of belief, is distinct from hope and knowledge.

I have spoken of God and immortality as twin objects of faith, and later will return to this linkage. Now let me say only that when I use the word *God* it means a God who has provided for our survival after death. When I use the word *immortality* it means survival in whatever manner God has provided. Following such fideists as Immanuel Kant and Miguel de Unamuno, and in line with the overwhelming majority of theists past and present, I will assume that the two beliefs go hand in hand and are mutually reinforcing.

Not that they can't be separated. Many thinkers have professed faith in God while denying an afterlife, but in almost every case the God involved is a pantheistic deity. A God more or less synonymous with Being or Nature obviously need not be concerned over whether we mortals live again, but such a God is the God of Spinoza, or the impersonal Absolute of Hegel and F. H. Bradley, not a God modeled on human personality. A personal God who did not provide for immortality would be a God less just and merciful than you and I. The whole point of the person model is to elevate human attributes, not lower them.

It is easier, perhaps, to hope for or even believe in an afterlife without faith in a personal God. One simply regards survival as part of the nature of things.[3] This point of view is sometimes taken by atheists and pantheists who believe in reincarnation. Among modern philosophers, John Ellis McTaggart of Cambridge University and the French-born American Curt John Ducasse were notable in combining nonbelief in God with a belief in the preexistence and the afterlife of human souls. Robert Ingersoll, the famous American infidel, never hesitated to denounce any kind of deity, yet he was curiously open-minded about life after death. One may, of course, hope for immortality and at the same time estimate the odds against it as high, as one may hope to win a sweepstakes without believing the win is likely.

Although it is possible to believe in a personal God without believing in immortality, and vice versa, both views are extremely rare, and in any case they play no role in what follows. I agree with Unamuno that for almost all theists God is essentially the provider of immortality. Did any religious leader ever emphasize this more than Jesus? In the first chapter of *The Tragic Sense of Life*, Unamuno tells of suggesting to a peasant that there might be a God who governs heaven and earth, but that we may not be immortal in the traditional sense. The peasant responded, "Then wherefore God?"

Let us now inquire as to the sources of this faith. What prompts some men and women and not others to make that quixotic somersault of the soul, what Kierkegaard called the "leap of faith"? What enables them to turn themselves around, like Dante at the center of the earth when he began his climb from hell to purgatory, and believe in God and immortality even though both beliefs are unsupported by reason or science; even though both are plainly counterindicated by persuasive arguments?

Most people never worry about why they believe any religious doctrine. They just absorb their beliefs, often conflicting, from parents, relatives, friends, and surrounding cultures. But insofar as one is capable of deciding whether to believe or not believe in God and immortality, or at least to reflect about such a decision, what can be said to justify this leap?

Perhaps there is built into human nature a natural tendency toward faith, something comparable to a natural thirst for water. This is, of course, an ancient notion. In modern terms, is there a genetic basis for faith? Some sociobiologists have raised this possibility. Maybe it is balanced by a genetic predisposition toward atheism, like conflicting genetic impulses toward egoism and altruism. Maybe the relative strengths of the two tendencies vary with individuals, and vary statistically with cultures. I do not know the answers to these questions.

Assume there is no genetic basis for either atheism or altruism. The same questions return on an environmental basis. It is obvious that most cultures within recorded history have been dominated by religious systems. Can we say that in a reasonably healthy society, one that does a good job of meeting human needs, the healthier members of that society make leaps of faith? Even Freud, for whom religion was a neurosis, considered the possibility that all cultures need such illusions to remain happy and secure.

Soviet philosophy was officially atheist, and for decades we watched a remarkable religious experiment in Russia—an attempt by the government to stamp out religious faith among its citizens. How successful was this? Even among Soviet officials, were the majority genuine atheists or were they closet believers? Did the great Soviet campaign for atheism influence Russia for good or ill? Was there a direct or inverse correlation in Russia, or for that matter anywhere else, between mental health and faith? I do not know the answers to these questions.

There is, however, one fact about which both atheists and theists can agree. For many people, perhaps most people, there is a deep, ineradicable desire not to cease to exist. Perhaps this desire, this fear of falling into what Lord Dunsany once called the "unreverberate blackness of the abyss," is no more than an expression of genetic mechanisms for avoiding death. Or is it more? It is easy to understand why any person would think death final—everything in our experience indicates it— but I share with Unamuno a vast incredulity when I meet individuals, seeming well adjusted and happy, who solemnly assure me they have absolutely no desire to live again. Do they really mean it? Or are they wearing a mask which they suppose fashionable while deep inside their hearts, in the middle of the night and in moments of agony, they secretly hope to be surprised someday by the existence and mercy of God?

That the leap of faith springs from passionate hope and longing or, to say the same thing, from passionate despair and fear, is readily admitted by most fideists, certainly by me and by the fideists I admire. Faith is an expression of feeling, of emotion, not of reason. But, you may say, does not this lower faith? Is not man the only rational animal? No. Emotion more than reason, certainly as much as reason, distinguishes us from the beasts. "More often I have seen a cat reason," wrote Unamuno, in that marvelous chapter to which I referred a moment ago, "than laugh or weep." Yes, and I have watched my desk calculator reason more often than laugh or weep.

Freud thought of faith as little more than a desire to obtain in one's adult life the warmth, security, and comfort of the child who is cared for by loving parents. Of course! What else? Friedrich Schleiermacher said it all when he spoke of faith as springing from our feeling of "creaturehood," our dependence on outside help for our survival.

The true fideist grants it all. He may—in my opinion, should—go even another step, the ultimate step, in conceding points to the atheist.

Not only are there no compelling proofs of God or an afterlife, but our experience strongly tells us that Nature does not care a fig about the fate of the entire human race, that death plunges each of us back into the nothingness that preceded our birth. Is there need to elaborate the obvious? Thousands of good people are killed by an earthquake. Where is God? Not only is there no God, said Woody Allen, but try getting a plumber on weekends. So dependent is the mind on the material structure of the brain that genetic damage, drugs, injuries, diseases, operations, and senility can severely alter one's personality and ability to think and act normally. Even ordinary sleep can wipe out consciousness. If there is a soul capable of existing apart from that gray lump of tissue inside every skull, it is as hidden from us as God is hidden.

I agree with Pierre Bayle and with Unamuno that when cold reason contemplates the world it finds not only an absence of God, but good reasons for supposing there is no God at all. From this perspective, from what Unamuno called the "tragic sense of life," from this despair, faith comes to the rescue, not only as something nonrational but in a sense irrational. For Unamuno the great symbol of a person of faith was his Spanish hero Don Quixote. Faith is indeed quixotic. It is absurd. Let us admit it. Let us concede everything! To a rational mind the world *looks* like a world without God. It *looks* like a world with no hope for another life. To think otherwise, to believe in spite of appearances, is surely a kind of madness. The atheist sees clearly that windmills are in fact only windmills, that Dulcinea is just a poor country bumpkin with a homely face and an unpleasant smell. The atheist is Sarah, justifiably laughing in her old age at Abraham's belief that God will give them a son.

What can be said in reply? How can a fideist admit that faith is a kind of madness, a dream fed by passionate desire, and yet maintain that one is not mad to make the leap?

Persons of strong faith sometimes say they have a direct awareness of God, a knowledge of the sort that philosophers have called "knowledge by acquaintance." Mystics claim to have perceived God in a manner analogous to looking at the sun. We shall not linger over these claims. They carry no weight with anyone who has not had such an experience. No empirical tests can confirm that a person who professes such contact with God is actually in such contact. In many cases of persons who claimed such visions there is good evidence that they were experiencing delusions.

A subtler argument was made famous by Kant. Pure reason, said Kant, can prove neither God nor immortality, nor can it show them to be impossible. But we do not live by pure reason alone. We also live by what Kant called practical reason, by what a modern Kantian could call pragmatic reason. Everyone, said Kant, has a sense of duty, a conscience (Freud's superego). It tells us there is a difference between right and wrong, that it is our duty to be as good as we can and thereby promote the *summum bonum*, the highest good for humanity. This "moral law" within us is so powerful and awesome, as awesome as the spectacle of the starry heavens,[4] that we cannot escape believing that the highest good will someday be realized.

But look around. You see virtuous people, often children, suffering and dying for no apparent reason. At the same time you see wicked persons living healthy, happy, and prosperous lives until they die peacefully of old age. Where is the justice in such a scene? It can be just, said Kant, only if we assume another life, a life in which good is rewarded and wickedness punished. Not only that, but the perfection of goodness, for every individual, demands unlimited time in which to grow and profit from experience. Our life is cut off when we have just started to learn how to live. If there is no afterlife, no future in which virtue and happiness can be correlated, then our sense of morality becomes a sham. It arouses in us a passionate hope that can never be fulfilled.

Kant did not regard these arguments as proofs in the sense that one can prove a theorem in mathematics or establish a fact or law of science with high probability. We cannot "know" there is an afterlife. All Kant insists on is this. If we take seriously our hope that justice will be done with respect to our lives, we must posit an afterlife. And if there is an afterlife, there must be a God who is good enough and powerful enough to provide it. It is not our *duty* to believe in God and immortality. Our duty is only to be good, and many atheists (Kant singled out Spinoza) can be very good. But if we want to make our beliefs consistent with the demands of our moral nature, we must posit God and immortality. And if we have faith, we do more than recognize them as posits. We also believe them to be true.

What should a modern fideist make of this? I think there is much to be said for Kant's arguments, and I will return to them again, but I agree with Unamuno that behind the complicated language of Kant's *Critique of Practical Reason* is one simple fact that Kant did not fully admit even to himself. As a man of flesh and bone, to use one of Unamuno's favorite expressions, Kant passionately desired God and immortality. He may have thought he posited God to make sense of morality; actually he posited God because he needed God in order to live. But let us listen to Unamuno himself, as he writes about Kant in the first chapter of *The Tragic Sense of Life*:

Take Kant, the man Immanuel Kant, who was born and lived at Königsberg, in the latter part of the eighteenth century and the beginning of the nineteenth. In the philosophy of this man Kant, a man of heart and head—that is to say, a man—there is a significant somersault, as Kierkegaard, another man—and what a man—would have said, the somersault from the *Critique of Pure Reason* to the *Critique of Practical Reason*. He reconstructs in the latter what he destroyed in the former, in spite of what those may say who do not see the man himself . . .

Kant reconstructed with the heart that which with the head he had overthrown. And we know, from the testimony of those who knew him and from his testimony in his letters and private declarations, that the man Kant, the more or less selfish old bachelor who professed philosophy at Könisberg at the end of the century of the Encyclopedia and the goddess of Reason, was a man much preoccupied with the problem—I mean with the only real vital problem, the problem that strikes at the very root of our being, the problem of our individual and personal destiny, of the immortality of the soul. The man Kant was not resigned to die utterly. And because he was not resigned to die utterly he made that leap, that immortal somersault, from the one Critique to the other.

Whosoever reads the *Critique of Practical Reason* carefully and without blinkers will see that, in strict fact, the existence of God is therein deduced from the immortality of the soul, and not the immortality of the soul from the existence of God. The categorical imperative leads us to a moral postulate which necessitates in its turn, in the teleological or rather eschatological order, the immortality of the soul, and in order to sustain this immortality God is introduced. All the rest is the jugglery of the professional of philosophy.

Kant argued that it was necessary to posit God to satisfy a universal human desire for moral justice. Unamuno did not disagree. He simply saw more clearly than Kant, or perhaps more clearly than Kant was willing to admit, that the desire for moral justice flows from a deeper passion. For Unamuno, for all those who do not want to die, who do not want those whom they love to die, God is a necessary posit to escape from unbearable anguish. It is easy to say with the head that God does not exist, but to say it with the heart? "Not to believe that there is a God or to believe that there is not a God," wrote Unamuno, "is one thing; to resign oneself to there not being a God is another thing, and it is a terrible and inhuman thing; but not to wish that there be a God exceeds every other moral monstrosity; although, as a matter of fact, those who deny God deny Him because of their despair at not finding Him."[5]

Psalm 14:1, Unamuno liked to remind his readers, does not say, "The fool hath said in his *head*, There is no God."

There is another way to approach the task of justifying faith. I like to view it as a generalization of Blaise Pascal's famous "wager," but first let us see how Pascal himself presented it. Pascal was a Roman Catholic, and like all Catholics of his day he believed that every human soul had one of two destinies: eternal happiness in heaven or eternal misery in hell. Moreover, he believed that the soul's future state depended on accepting or rejecting Catholic doctrine. A person knowing of the

Church's claim is thus faced with two alternatives. He may accept or reject the Church. In either case, the Church's doctrines may be true or false. Suppose he accepts. The payoff is infinite happiness if the Church is right; at the most a finite loss if it is wrong. Suppose he rejects. The payoff is at most a finite gain if the Church is wrong, but infinite misery if the Church is right. In view of these possibilities, said Pascal, is not joining the Church clearly the best bet?[6]

It is hard to imagine a reader of this book being impressed by Pascal's argument. As numerous critics have pointed out, even in Pascal's time the Muslim religion offered the same monstrous alternatives to potential converts. But if you wagered on immortality in the Islamic heaven you ran the risk of misery in the Christian hell. Who could genuinely convert to all religions that offered similar alternatives? Nevertheless, behind Pascal's wager there is a broader notion that can be applied to belief in God and immortality quite apart from the salvation doctrines of any organized church. This generalized Pascalian wager is suggested as far back as Plato's *Phaedo*, where Socrates, before drinking hemlock, speaks of belief in another life as a worthy risk because "fair is the prize and the hope great." In recent times the generalized wager has found its classic defense in William James's "The Will to Believe," an essay to which I now turn.

James argued his case with more care than he is usually credited with. First he makes a distinction, essential to all that follows, between what he calls a live option and a dead or trivial option. A live option is a choice between two alternatives that meets three provisos:

1. The alternatives must be plausible enough so that you are truly capable of deciding either way. Should you believe the earth flat or round? This was once a live option. Today it is dead, except perhaps for a handful of ignorant Protestant fundamentalists. Should you believe or not believe that the Reverend Sun Myung Moon is the new Messiah? This is a live option for some naive young people, but for most people it is not. Should you spend your next vacation in Indianapolis? This is not a live option if you have no reason or desire to go there.

2. The choice must be forced. It must be what Kierkegaard liked to call an either/or. James gives the humble counterexample of choosing between going out with or without an umbrella. You can avoid the choice by not going out at all. Should you become a Scientologist or a Moonie? Clearly you need not join either cult. But the choice between voting for candidate X or not voting for candidate X is an either/or.

3. The alternatives must be momentous, not trivial. Should you have an egg for breakfast? This meets the first two criteria, but it is not a live option because the alternatives are too unimportant. Should you marry a certain person? Now the question is not so trivial.

James's thesis can be put simply. When we are confronted with live options, and when there are insufficient grounds for deciding rationally, we have no other way to decide except emotionally, by what James called our "passional nature." Who can deny that when a momentous decision is thrust upon us, and the head cannot decide, the heart must take over? But James is saying more than that. He is saying that there is nothing irrational or absurd about letting the heart take over.

The question "Does God exist?" confronts many people, perhaps most people, as a live option. The choice is forced in the sense that one must either believe or not believe. The agnostic may not insist there is no God, but he has exercised his option not to believe. Elsewhere, James likens the agnostic to a person who refuses to stop a murder, or to bail water from a sinking ship, or to save one's life by leaping across a chasm. The metaphors are overdramatic and pejorative but the basic point is sound. To avoid making an emotional decision about a live option (when there are no grounds for a decision) is itself an emotional decision, and one that can have momentous consequences. Like Kierkegaard, James spoke of faith as a "leap in the dark." A leap across a precipice is made at our peril. Of course one would be foolish to make such a leap for no reason at all, but if there are no reasons, then it is not a live option. The decision to believe or not believe in God, James maintained, is for many people a live option because it makes a difference in how they feel and how they live.

The leap of faith is made at our peril, yes, but so is the decision not to leap. James expressed it polytheistically. A man who shuts "himself up in snarling logicality," and demands that "the gods extort his recognition willy-nilly, or not get it at all," may be cutting "himself off forever from his only opportunity of making the gods' acquaintance."

James closes his essay with a passage from Fitzjames Stephen:

> We stand on a mountain pass in the midst of whirling snow and blinding mist, through which we get glimpses now and then of paths which may be deceptive. If we stand still we shall be frozen to death. If we take the wrong road we shall be dashed to pieces. We do not certainly know whether there is any right one. What must we do? "Be strong and of a good courage." Act for the best, hope for the best, and take what comes . . . If death ends all, we cannot meet death better.

The resemblance to Pascal's wager is obvious, and in other writings, especially in his essay on "The Sentiment of Rationality," James makes the parallel even stronger. In substance this is what he says: Belief in God and immortality are unsupported by logic or science, but because they are live options we cannot avoid an emotional decision. If for you the leap of faith makes you happier, then for you faith is the best bet. You have much to gain and little to lose. You have a *right* to believe. (In later years James said he should have called his essay "The Right to Believe.") I think James would have liked the way Count Manuel, in James Branch Cabell's *Figures of Earth*, formulates the wager:

> "That may very well be, sir, but it is much more comfortable to live with than your opinion, and living is my occupation just now. Dying I shall attend to in its due turn, and, of the two, my opinion is the more pleasant to die with. And thereafter, if your opinion be right, I shall never even know that my opinion was wrong: so that I have everything to gain, in the way of pleasurable anticipations anyhow, and nothing

whatever to lose, by clinging to the foolish fond old faith which my
fathers had before me," said Manuel, as sturdily as ever.[7]

It was characteristic of James, as we saw in earlier chapters, to indulge at times in rhetoric that made his views easy to ridicule. His
many examples of how "faith in a fact can help create the fact"—such
as belief in one's career, or in winning a battle or a football game—
surely have no relevance to faith in God. As if somehow believing in
God could help make God a reality! George Santayana, in his marvelous essay on James,[8] is right in attacking James for these excesses (a
product, in my opinion, of James's enthusiasm coupled with a confused
epistemology), but Bertrand Russell's attack (in the chapter on James
in his History of Western Philosophy) is as wide of the mark as his attack on Dewey in the chapter that follows. Russell actually suggests
that James's arguments would establish the truth of "Santa Claus exists" as readily as "God exists," although it should have been clear to
Russell that the Santa Claus hypothesis is not a live option for anyone
except a young child. James might well have argued that under certain
conditions a child is justified in believing in Santa Claus, but as soon
as the child matures enough to understand the evidence against the
hypothesis, the belief tips the other way.

On the other hand, I think Russell is right in saying that James
often wrote as if he were concerned only with the pragmatic consequences of thinking and acting as if God exists, not with the question
of whether God actually does exist. In my opinion James did not regard
faith in such a superficial way. In any case it is not how most fideists
regard faith.

It should be obvious that anyone who manages the leap of faith
does not say to himself or herself: "I really don't know whether God
exists or not, or whether there is another life, but because I find these
beliefs comforting I shall pretend they are true." Perhaps some philosophers have been capable of this crazy "as if" approach to faith, but I
have never met a theist who thought that way. Quite the contrary! For
a person of faith, belief in God's reality is usually stronger than belief
in any scientific hypothesis.

This was true even of Kant. It surely is a mistake to accuse Kant of the *als ob* perversion that some of his followers proclaimed in Germany. True, belief in God is not knowledge, but Kant, as he himself said, denied knowledge in order to make room for faith. For Kant, as for Plato, the phenomenal world, the phaneron open to exploration by science, is less real than the transcendent world of which the phaneron is but a shadow. It is our transcendent self, momentarily trapped in space-time, that believes by faith in a transcendent God. Not only does Kant avoid the notion of God as an "as if" posit less real than the universe, he argues the exact opposite.

Let me speak personally. By the grace of God I managed the leap when I was in my teens. For me, faith was once bound up with an ugly Protestant fundamentalism. I outgrew this slowly, and eventually decided I could not even call myself a Christian without using language deceptively, but faith in God and immortality remained. Much of my novel, *The Flight of Peter Fromm*, reflects these painful changes. The original leap was not a sharp transition. For most believers there is not even a transition. They simply grow up accepting the religion of their parents, whatever it is. For others, as we all know, belief can come suddenly, in an explosive conversion experience as startling as a thunderclap.

James applied the term "over-beliefs" to beliefs supported only by the heart. This does not mean they are not genuine beliefs. It only means they are beliefs of a special and peculiar sort. Why do I believe in the Pythagorean theorem? Because I can follow a deductive proof that rests on the posits of Euclidean geometry, and because the theorem is confirmed by experience. But this is not a choice about a live option. In a sense, belief in the formal truth of the theorem is a trivial, empty belief. It tells me only that if I accept certain posits, and rules for manipulating certain symbols, I am allowed to form a chain of those symbols that can be interpreted as the Pythagorean theorem. I believe in the formal truth of the theorem for much the same reason that I believe no bachelor is married.

Mathematical theorems are useful because they apply to the physical world, but (as I have said earlier) the applications require what

Rudolf Carnap called correspondence rules, such as identifying the number 1 with a pebble or a star, or a straight line with a ray of light. As soon as we move from pure mathematics to applied mathematics, we move to a realm where hypotheses become uncertain, where the best we can do is weight them with varying degrees of credibility. Naturally we believe most strongly in those assertions of science that seem to us the best confirmed, but belief in God can carry with it a certitude, springing from the heart, that is stronger than any belief about the world. It is easier for me to believe that any fact or law of science is no more than a momentary illusion, produced by the Great Magician and subject to change whenever the Great Magician decides to modify his Act, than to believe that the Great Magician doesn't exist.[9] But this certainty is not knowledge of the kind we have in mathematics or science. It is trivially true that we believe what we know, or think we know. To believe what we do not know, what we hope for but cannot see—this is the very essence of faith.

I am quite content to confess with Unamuno that I have no basis whatever for my belief in God other than a passionate longing that God exists and that I and others will not cease to exist. Because I believe with my heart that God upholds all things, it follows that I believe that my leap of faith, in a way beyond my comprehension, is God outside of me asking and wanting me to believe, and God within me responding. This has been said thousands of times before by theists. Let us listen to how Unamuno says it:

> Wishing that God may exist, and acting and feeling as if He did exist, and desiring God's existence and acting conformably with this desire, is the means whereby we create God—that is, whereby God creates Himself in us, manifests Himself to us, opens and reveals Himself to us. For God goes out to meet him who seeks Him with love and by love, and hides Himself from him who searches for Him with the cold and loveless reason. God wills that the heart should have rest, but not the head, reversing the order of the physical life in which the head sleeps and rests at times while the heart wakes and works unceasingly. And thus knowledge without love leads us away from God; and

love, even without knowledge, and perhaps better without it, leads us to God, and through God to wisdom. Blessed are the pure in heart, for they shall see God![10]

No more can be said. The leap of faith, in its inner nature, remains opaque. I understand it as little as I understand the essence of a photon. Any of the elements I listed earlier as possible causes of belief, along with others I failed to list, may be involved in God's way of prompting the leap. I do not know, I do not know! At the beginning of the leap, as at the beginning of all decisions, is the mystery of free will, a mystery that for me is inseparable from the mysteries of time and causality, and the mystery of the will of God.

22. THE COLOURED LANDS

Gilbert Keith Chesterton, or G.K., as he was often called, was quite a good artist. He illustrated his own book of short stories, *The Club of Queer Trades* (the 1988 Dover paperback edition of which includes all this art), and six books by his friend Hilaire Belloc. His unpublished illustrations for Wilkie Collins's mystery novel *The Moonstone* are in the John Shaw collection of Chesterton material owned by the University of Notre Dame.

The Coloured Lands is as far as I know the only book that contains a selection of Chesterton drawings in full color. What follows here is my introduction to Dover's 2008 reprint.

"The amazing thing about the universe," Chesterton writes in his chapter on wonder and the wooden post, "is that it exists." The amazing thing about G. K. Chesterton is that *he* existed.

If G. K. Chesterton is one of your heroes, as he is one of mine, you'll find *The Coloured Lands* crammed with some of G.K.'s charming youthful writings. The book was put together by Maisie Ward in 1938, two years after Chesterton's death. Ward was G.K.'s friend and biographer. Some dozen lives of Chesterton have since been written, none better or more accurate than Ward's 685-page *Gilbert Keith Chesterton*.

"The Coloured Lands," a short story in the book of the same name, is about a strange young man who lets a boy named Tommy look through four pairs of spectacles made with colored glass that turn everything blue, red, yellow, or green. The man tells Tommy that when he was a child he had been fascinated by colored glasses, but soon tired of seeing the world in single colors. In a rose red city, he explains, you can't see the color of a rose because everything is red. At the suggestion of a powerful wizard, the man is told to paint the scenery any way he liked:

"So I set to work very carefully; first blocking in a great deal of blue, because I thought it would throw up a sort of square of white in the middle; and then I thought a fringe of a sort of dead gold would look well along the top of the white; and I spilt some green at the bottom of it. As for red, I had already found out the secret about red. You have to have a very little of it to make a lot of it. So I just made a row of little blobs of bright red on the white just above the green; and as I went on working at the details, I slowly discovered what I was doing; which is what very few people ever discover in this world. I found I had put back, bit by bit, the whole of that picture over there in front of us. I had made that white cottage with the thatch and that summer sky behind it and that green lawn below; and the row of the red flowers just as you see them now. That is how they come to be there. I thought you might be interested to know it."

As far as I know, Chesterton never read any of L. Frank Baum's fourteen Oz books. Had he done so, I believe he would have been entranced to learn that each of the five regions of Oz is dominated by a single color. Munchkin land is blue, the Winkie region is yellow, the southern Quadling country is red, the northern Gillikin region purple, and of course the Emerald City is green. Chesterton was fond of every color. One of his best essays, "The Glory of Gray," praises gray for its power as background to intensify other colors.

In her fine introduction to *The Coloured Lands*, Maisie Ward reminds us of what G.K. called the "central idea" of his life—that we should all learn to see everything, from a sunflower to the cosmos, as a miracle, and to be perpetually thankful for the privilege of being alive. Would things not be much simpler and easier to comprehend if *nothing* existed? This sense of awe toward the terrible mystery of why there is something rather than nothing—of why, as Steven Hawking once wrote, the universe "bothers to exist"—can arouse an emotion so close to what Sartre called nausea, and William James called a "metaphysical wonder sickness," that if it lasted more than a few moments we could go mad. In *The Coloured Lands* this emotion comes through strongly in two chapters, one an essay on "Wonder and the Wooden Post," the other a short story titled "Homesick at Home."

Here is a passage from "Wonder and the Wooden Post":

When the modern mystics said they liked to see a post, they meant they liked to imagine it. They were better poets than I; and they imagined it as soon as they saw it. Now I might see a post long before I had imagined it—and (as I have already described) I might feel it before I saw it. To me the post is wonderful because it is *there*; there whether I like it or not. I was struck silly by a post, but if I were struck blind by a thunderbolt, the post would still be there; the substance of things not seen. For the amazing thing about the universe is that it exists; not that we can discuss its existence. All real spirituality is a testimony to this world as much as the other: the material universe does exist. The Cosmos still quivers to its topmost star from that great kick that Dr. Johnson gave the stone when he defied Berkeley. The kick was not philosophy—but it was religion.

"Homesick at Home" anticipates Chesterton's long-forgotten novel *Manalive*, back in print in 2000 as a Dover paperback. In this short story a man with the unlikely name of White Wynd sets out on a trip around the world so he can return to recapture a fading sense of wonder and gratitude for his wife and home. In *Manalive*, one of G.K.'s funniest fantasies, White Wynd is replaced by Innocent Smith, who circumnavigates the globe for the same reason Wynd did. I urge you to read this curious novel. You'll find a chapter about it in my book *The Fantastic Fiction of Gilbert Chesterton*.

One of G.K.'s finest poems, "A Second Childhood," concerns his celebrated sense of wonder toward a universe that to an atheist is what Chesterton called, in an essay on Shelley, "the most exquisite masterpiece ever constructed by nobody." I think it is one of the greatest religious poems ever written. Allow me to quote it in full:

When all my days are ending
And I have no song to sing,
I think I shall not be too old
To stare at everything;
As I stared once at a nursery door
Or a tall tree and a swing.

Wherein God's ponderous mercy hangs
On all my sins and me,
Because He does not take away
The terror from the tree
And stones still shine along the road
That are and cannot be.

Men grow too old for love, my love,
Men grow too old for wine,
But I shall not grow too old to see
Unearthly daylight shine,
Changing my chamber's dust to snow
Till I doubt if it be mine.

Behold, the crowning mercies melt,
The first surprises stay;
And in my dross is dropped a gift
For which I dare not pray:
That a man grow used to grief and joy
But not to night and day.

Men grow too old for love, my love,
Men grow too old for lies;
But I shall not grow too old to see

Enormous night arise.
A cloud that is larger than the world
And a monster made of eyes.

Nor am I worthy to unloose
The latchet of my shoe;
Or shake the dust from off my feet
Or the staff that bears me through
On ground that is too good to last,
Too solid to be true.

Men grow too old to woo, my love,
Men grow too old to wed:
But I shall not grow too old to see
Hung crazily overhead
Incredible rafters when I wake
And find I am not dead.

A thrill of thunder in my hair;
Though blackening clouds be plain,
Still I am stung and startled
By the first drop of the rain:
Romance and pride and passion pass
And these are what remain.

Strange crawling carpets of the grass,
Wide windows of the sky:
So in this perilous grace of God
With all my sins go I:
And things grow new though I grow old,
Though I grow old and die.

23. THE PRICE WE PAY

What? Shall we receive good from the hand of God, and shall we not receive evil?

—Job 2:10

The following essay, disguised as a book review, ran in *The New Criterion*, November 2008.

BART D. EHRMAN

GOD'S PROBLEM: HOW THE BIBLE FAILS TO ANSWER OUR MOST
IMPORTANT QUESTION—WHY WE SUFFER.
HARPERONE, 304 PAGES, $25.95

I have just finished reading *God's Problem* by Bart D. Ehrman, a professor of religious studies at the University of North Carolina at Chapel Hill. His earlier book, *Misquoting Jesus*, made *The New York Times* bestseller list. A former fundamentalist, Ehrman graduated from the Moody Bible Institute in Chicago, did graduate work at Wheaton College (Billy Graham's alma mater), and obtained a doctorate at Princeton Theological Seminary. Slowly over the years, he lost his faith in Christianity. His new book explains why. It is the latest in

a surprising spate of books defending atheism. The book's subtitle is
*How the Bible Fails to Answer Our Most Important Question—Why
We Suffer.*

Like all writers on the topic (theist, atheist, or pantheist), Ehrman
distinguishes two main aspects of the so-called problem of evil:

1. Evils caused by human behavior. A demented man fires an automatic into a crowd. The lives of those killed are as irrationally
   ended as if they had been killed by an earthquake. Hitler murders millions of Jews. Stalin murders even more without regard
   for race, color, or creed.
2. Evils caused by nature.

Christian theologians, going back to St. Augustine and earlier, have
reasoned that God is unwilling to prevent such crimes by withholding
his gift of free will. If we lacked free will, Gilbert Chesterton liked to
say, there is no point in thanking someone for passing the mustard.
Free will is at the heart of human consciousness. We can't have one
without the other. We are not robots doing what we are wired to do by
heredity and experience. But if we are free to do good or evil, so goes
the argument, our very freedom makes evil behavior possible. If it were
otherwise, the earth would be populated not by humans but by robotic
featherless bipeds similar to the social insects—bees, wasps, and ter-
mites. This is a plausible argument, and Ehrman does a good job of
presenting it even though he doesn't buy it.

Why God permits natural evil is not so easy to explain. An earthquake can end the lives of thousands. Millions in Africa may die of
starvation. In Genesis we read about a flood that drowns almost the
entire human race, including little babies. In that case the murderer
was not nature but God himself. I once had lunch with a fundamentalist Seventh-day Adventist. When I asked him how he defended God's
drowning of innocent infants he astonished me by saying that God
foresaw the future and knew that the babies would all grow up to become malevolent men and women! I was tempted to stand and shout
"Touché!" It was a thought that had never occurred to me.

Why does God permit massive suffering? An old argument—it traces back to ancient Greece—goes as follows. God is either incapable of abolishing natural evil, in which case he is not omnipotent, or he can but won't, in which case he is not good. How can a theist go between the horns of this dreadful dilemma?

Of course this is no problem for an atheist. Evils are simply the way the world is. But for a theist the problem can be agonizing. Indeed, it is probably why most atheists are atheists. There is an answer, though not one likely to persuade any atheist. Surprisingly, Ehrman only briefly mentions it in connection to Rabbi Harold Kushner's popular 1981 book *Why Bad Things Happen to Good People*.

Kushner's argument rests on the belief of many theists, past and present, that there are severe limits on the powers of any sort of deity. Thomas Aquinas somewhere writes that there are many things God cannot do. One, he can't alter the past. I doubt if anyone today thinks God could, if he liked, erase Hitler from history. Two, God cannot do things that are logically impossible. The saint's example: God can't create a perfect human who at the same time is a perfect horse. A mathematician can add that God can't make a triangle with four sides, or cause 2 plus 2 to equal, say, 7.

Not only must pure mathematics be free of logical contradictions, but applied mathematics as well. If objects in the outside world maintain their identities, then two apples plus two more apples can't result in any number of apples except four. The same is true of cows, stars, and all other things that model the number 1. It is best, Aquinas wrote, not to say there are things God can't do, but to say there are things that can't be done.

Let's see how this applies to natural evils. When God created the universe, or as a theist would say, started the process of creation, he not only limited the process to a world free of contradictions, the world also had to obey unalterable laws. It is not possible, say, for planets to go around the sun in elliptical orbits, and at the same time travel in square orbits. It is necessary also that gravity remain constant. Life could not exist if gravity turned into a repulsive force that sent everything flying off into space. If the earth suddenly stopped rotating, as the Bible's account of Joshua's miracle suggests, the result would be equally catastrophic. Indeed, if all laws were not unbreakable the world would be far too chaotic to support life.

The fact that stable laws are essential for any conceivable universe with sentient life at once makes natural evils inevitable. If someone carelessly loses balance at the edge of a cliff and topples over, you can't expect God to suspend gravity in the region and allow the person to float gently down. If a piece of heavy masonry dislodges from the top of a tall building, and is on its way toward the head of someone on the sidewalk, you can't expect God to divert its path or turn it into feathers.

Suppose a man falls asleep while driving a car down a thruway. He crosses the median and smashes into another car, killing a woman and her three children. Such tragedies are the terrible price we pay for a universe with unalterable laws of velocity and momentum. If God were obliged to prevent all accidents that kill or injure, he would have to be constantly poking his fingers into millions of events around the globe. History would turn into a chaos of endless miracles.

The necessity of order in the universe can also explain why God doesn't intervene to prevent medical horrors. Consider the Black Death that killed a third of Europe's population. Why did God not prevent this awful plague? A possible answer, weak though it may seem, is that the existence of deadly microbes was the inevitable consequence of biological laws essential to the evolution of intelligent creatures. From this perspective, evolution was perhaps the only way God could fabricate such unlikely animals as you and me. Irrational deaths from diseases and other biological causes such as cancer are the prices we pay for evolution—for the miracle of being alive.

It is easy to see how similar arguments apply to natural disasters such as earthquakes, tsunamis, volcanic eruptions, lightning that starts fatal fires, and other natural evils. Laws of physics obviously apply to movements of the earth's crust that cause earthquakes. Laws of rain and lightning make inevitable the occasional starting of fires. Deaths from quakes and lightning are the prices we pay for the laws of physics, without which there could be no universe. Do I *know* this is why God permits such disasters? I do not. I only put the explanation forward (it goes back to Maimonides and even earlier) as the best I have encountered in the vast literature on the topic.

The most famous defense of this explanation was the *Theodicy*, written by the great German mathematician and philosopher Gottfried Wilhelm Leibniz. He coined the term "compossible" for a universe free of logical contradictions and with unvarying laws. He imagined God considering all compossible universes, each with its unique set of laws. Many physicists today, especially those working on superstring theory, not only take seriously Leibniz's vision of a "multiverse" containing perhaps an infinity of compossible universes, they believe such a multiverse actually exists!

Leibniz further imagined that God selected for creation the universe with the smallest amount of unavoidable human misery. The notion that Leibniz was naively unaware of the vast amount of pain in our world makes him out to be an idiot, which of course he wasn't. He even shared with Newton the invention of calculus! Voltaire's much admired satire about Dr. Pangloss missed the whole point of Leibniz's *Theodicy*. Suffering, for Leibniz, was the price we pay for a possible universe.

Leibniz also knew that humanity is capable of eliminating most irrational suffering. We can invent clever ways to construct buildings and houses that withstand earthquakes. We can find ways to prevent deaths from floods. Science can discover cures for the ills of both body and mind. Today we have vaccines that prevent polio and smallpox. We can construct artificial limbs. Blindness can be prevented by removing cataracts. Some fine day we may even find ways to forestall famines, and eliminate epidemics and wars.

Leibniz's vision can be given a contemporary form as follows. After God selected the best compossible universe—the one with the least amount of necessary suffering—he adopted what could have been the only way to create such a universe. Somewhere in a higher space he started a quantum fluctuation that triggered what astronomer Fred Hoyle derisively called a "Big Bang." The bang generated a set of fundamental particles, fields, and laws—a fantastic mix in which you and I were there *in potentia*. The particles and fields, together with a set of laws, were such that after billions of years gravity would form galaxies, the suns and planets, and on at least one small planet life would begin and ultimately evolve such grotesque creatures as you and me. History would begin its slow and painful crawl toward a utopia in which pain would be minimized. Humans would eventually, as H. G. Wells wrote at the end of his *Outline of History*, stand on the earth as on a footstool and stretch out their arms to the stars. Manifestly there is nothing new about this scenario. You find it in the writings of eminent theologians of all faiths, as well as in secular variants in which God plays no role.

Meanwhile, as the plot (God's or otherwise) unrolls, there is no denying that enormous evils, with their inevitable injustices, haunt human history. Millions still perish and suffer needlessly from earthquakes, accidents, disease, and other causes. Good people die young while bad people live comfortably to old age. Is there any way a caring God, whose eye is on the sparrow, can rectify such obvious injustices? The only conceivable way is to arrange for some sort of afterlife. Every theist then faces the following trilemma:

1. God is unable to provide an afterlife, in which case his power seems unduly limited.
2. God can provide an afterlife but chooses not to, in which case his goodness is tarnished.
3. God is both able and willing to provide an afterlife.

Kant's *Critique of Practical Reason*, unread today by most philosophers and even by most theologians, is a vigorous defense of the third horn. Kant did not want to disappear. True, there are intelligent people who insist they have no desire to live again—H. G. Wells and Isaac Asimov, to mention two. I think they lied. Carl Sagan, another atheist, was more honest. He said it would be wonderful if he survived death, but he saw no evidence for such a hope. Woody Allen recently said he had no desire to live on in his films: "I just don't want to die." Boswell, in his life of Samuel Johnson, tells Johnson about a conversation with David Hume. "Hume said he had no desire to live again. He lies, said Johnson, as you will quickly discover if you hold a pistol to his breast." The great Spanish poet, novelist, and philosopher Miguel de Unamuno once asked a farmer if he believed it was possible that there is a God but no afterlife. The rustic responded, "Then wherefore God?"

A strange question now arises: If there is an afterlife, will it be in a world with free will and science such as to permit both kinds of evil? An East Indian would almost surely answer yes. As for me, I haven't the slightest idea. How could I possibly know?

Back to *God's Problem*, the book that triggered my long-winded speculations. It is hard to imagine how a better, more persuasive volume could be written on why irrational evil implies atheism. When you read a book on the topic by an orthodox Christian, such as C. S. Lewis's *The Problem of Pain*, or his *A Grief Observed*, about the death of his wife Joy from cancer, you sense Lewis's agony as he struggles to believe his own arguments. It is not only his pain that troubles Lewis, it is also his awareness of the enormous amount of suffering that continues to plague humanity. By contrast, there is little agony in Ehrman's book. There is only a huge relief over finally abandoning a youthful theism.

Ehrman's rhetoric, eloquent and powerful, differs from the rhetoric of other books on evil in that his central theme is this: nowhere does the Bible give a satisfactory answer to why a benevolent God would allow such massive misery. A "Scripture Index" at the back of the book lists almost two hundred Old Testament verses, and more than one hundred New Testament verses.

Ehrman's detailed analysis of the Book of Job is at the heart of his treatise. He makes clear that Job is a stitched-together hybrid of two documents by different authors. The first describes scenes in the land of Uz that alternate with scenes in heaven in which God and Satan argue about Job's faith. The second is a much longer section of poetry. The wealthy Job endures incredible God-caused blows that include the destruction of seven sons and three daughters, yet Job's faith in a loving God never wavers.

The moral of this much admired fantasy is simple. Irrational suffering is an impenetrable mystery. "God knows something you don't know," I once heard Oral Roberts say at a funeral in Tulsa. Who are you, the Lord shouts at Job from a whirlwind, to question the motives of the creator of the universe?

PART VII

POLITICS

24. IS *SOCIALISM* A DIRTY WORD?

I have long enjoyed writing letters to the editor even though I know that few will be printed. I was pleased when the morning paper in the Oklahoma town where I live published (October 28, 2008) the following letter. Obama's election a week later was, of course, a great tipping point in American history.

To the Editor, *The Norman Transcript*:

There are two kinds of socialism. Democratic socialism has what economists call a "mixed economy"—one in which strong government controls are combined with democracy and a free market. The term was stolen by the Russian Marxists, and by Hitler and Mussolini for their respective forms of tyranny.

Milton Friedman, America's most famous conservative economist, in his book "Free to Choose," considers America, since the days of FDR, to be a democratic socialist state, though less so than the democratic socialisms of Europe. Recently, McCain and his supporters have branded Obama a socialist because he wants to "spread wealth around." Of course McCain knows full well that the purpose of our progressive income tax is precisely to redistribute wealth.

In the appendix to "Free to Choose," Friedman lists the planks

224 / WHEN YOU WERE A TADPOLE AND I WAS A FISH

in the platform of Norman Thomas, our country's leading socialist, the last time he ran for president. Every plank is now accepted by almost every Republican. As Thomas liked to say, "Most Americans don't know the difference between socialism, communism and rheumatism."

MARTIN GARDNER

On November 11 the paper published the following rejoinder by a local reader:

Martin Gardner wrote a letter to the editor printed Oct. 28 titled "Is socialism a dirty word?" I recommend that anyone having questions about socialism read F. A. Hayek's "The Road to Serfdom."

Socialism has come to mean chiefly the extensive redistribution of income through taxation and the institutions of the welfare state. It is nothing less than a system of government operated slavery. It uses the power of government to force one individual to work for another individual without the two individuals having any contact or communication with each other. Socialism is the state running your life—it's the path to a totalitarian government.

Milton Friedman wrote, "The battle for freedom must be won over and over again." Socialists in this nation must be persuaded or defeated if they and we are to remain free people.

If it's true that "Most Americans don't know the difference between socialism, communism and rheumatism" this reflects a lack of education and is a great threat to our individual freedom. Look at Venezuela.

There is a totalitarian government that was voted into power.

SIDNEY T. HANNA

NORMAN

I responded on November 17:

Sidney Hanna's Nov. 11 letter, replying to my letter headed, "Is socialism a dirty word?" recommends that anyone wanting to understand socialism should read Hayek's book "The Road to Serfdom." He couldn't have picked a worse book. Hayek's incredible theme is that any movement toward more federal controls of a free market leads inevitably toward a slave state.

This prophecy has been thoroughly discredited by one simple fact. Industrialized nations that have adopted forms of democratic socialism, notably Denmark and Sweden, are stronger democracies than the U.S. Here only about half of eligible voters vote, in Sweden and Denmark the percentage is close to 90. There are no signs in England, France, and Germany, all three with more powerful governments than ours, of serfdom. As I said in my letter, Milton Friedman, America's top conservative economist, recognized that even the U.S. became democratic socialist under FDR.

Hanna cites Venezuela as a socialist nation that has lost its freedoms. But Venezuela is not a democratic socialist country. It's a Cuba-styled dictatorship. Hayek's book, though still a Bible for ultra-conservatives who haven't learned the difference between socialism and communism, has long been an obsolete, hopelessly flawed work. Only in America has *socialism* become a dirty word.

MARTIN GARDNER

Socialism is, of course, a hopelessly fuzzy word. If it means a democratic government forced to redistribute wealth to pay for what government has to do that can't be done by private enterprise, then all politicians who call themselves liberals or progressives are socialists. Unlike the political leaders in England and Europe, for reasons expressed by Thomas's quip about rheumatism, they cannot call themselves socialists.

Because I am only a journalist, I'm not afraid to say I'm a democratic socialist. No one more vigorously opposed Russian Marxism, especially in the form it took under Stalin (who murdered more citizens than Hitler), than the democratic socialists both inside and outside Russia. Democratic socialism has a distinguished, honorable history. In no way is it a "road to serfdom."

For Obama things won't be easy. He faces at home a recession that could slide into a depression, and abroad a fanatical Iran armed with nuclear weapons and a bitter hatred of Israel and the United States. Thank heaven we don't have a confused president, and a Pentecostal vice president who doubts evolution, believes in faith healing, and whose former pastor speaks in tongues. Will some intrepid reporter ask Sarah Palin when she expects Jesus to return to establish a new heaven on earth?

What happened in 2008 to our economy is not hard to understand. Even Alan Greenspan ended an unintelligible speech to Congress by admitting he had made a terrible mistake when he overestimated the free market's ability to take care of a downturn. The nation wobbled back to the days of Herbert Hoover, who famously said that the only way a government can handle a depression is to "let it blow itself out." Greenspan turned into Greedspan. CEOs grabbed or tried to grab their retirement bonuses of tens of millions of dollars.

I once heard Norman Thomas, in his old age, speak at a symposium in Manhattan. His voice was as clear and strong as ever, but he walked with great difficulty. As he shuffled his way slowly to the podium, he uttered two words that brought down the house: "Creeping socialism." The creep turned into a leap in the fall of 2008. Ironically, it was not Democratic liberals who caused the leap, but George Bush's conservative economic advisers! How many of the government's emergency controls will survive after the crisis passes? Greenspan doesn't know, and neither do you.

NOTES

6. WHY I AM NOT A PARANORMALIST

1. It is hard to believe, but just such a chaos was recommended by Paul Karl Feyerabend, a Vienna-born philosopher of science who taught at the University of California, Berkeley. Feyerabend believed there is no scientific method (in the sense that there are no rules that cannot be violated), and that competing theories are as incommensurable as different cultures. Because he recommended that state-supported schools be allowed to teach anything taxpayers wanted, including astrology, parapsychology, Hopi cosmology, creationism, voodoo, and ceremonial rain dances, he became the favorite philosopher for believers in the paranormal who read and understood him. For my opinion of Feyerabend's "epistemological anarchism," see my 1982 article "Science and Feyerabend," reprinted in *Order and Surprise* (1983).
2. Samuel Goudsmit, *ALSOS* (1947)
3. Sociologists William Sims Bainbridge and Rodney Stark, writing on "Superstitions: Old and New" in *The Skeptical Inquirer* (Summer 1980), reported on their surveys of how beliefs in certain aspects of the current occult mania correlate with religious faith. They found that people with no professed religion were the most inclined to believe in ESP and extraterrestrial UFOs. Paranormal cults were strongest in areas where traditional churches were weakest.
4. *Fads and Fallacies in the Name of Science* (1952).
5. *Science: Good, Bad and Bogus* (1981).
6. Books and articles on modern pseudoscience are now far too numerous to cite. Readers interested in keeping up with current trends in bogus science are urged

to subscribe to *The Skeptical Inquirer*, edited by Kendrick Frazier, and *The Skeptic*, edited by Michael Shermer.

7. *Fads and Fallacies* (see note 4, above).

8. Joseph Banks Rhine, *The Reach of the Mind* (Winchester, MA: William Sloane Associates, 1947), chapter 11.

9. Schmidt's cockroach experiment is described in "PK Experiments with Animals as Subjects," *The Journal of Parapsychology* 34: 255–61, 1970. For some unintended humor, see the accounts of Schmidt's animal PK tests in *Psi—What Is It?* (New York: Harper & Row, 1975), by Louisa E. Rhine, and the chapter on "ESP and Animals," in *What's New in ESP* (New York: Pyramid Books, 1976), by Martin Ebon.

  When Walter J. Levy, Jr., was working for Rhine, he discovered that live chicken eggs could influence a randomizer so that it kept them warmer than chance would have otherwise permitted. Schmidt had earlier reported on similar experiments with his cat. For Levy's work on fertilized eggs see "Possible PK by Young Chickens to Obtain Warmth," by Levy and E. André, in *The Journal of Parapsychology* 34, 1970, p. 303. Because Levy was later caught cheating while testing the ability of rats to influence a randomizer (see note 13, below), references to his pioneering work on egg PK have dropped out of psi literature.

10. The quotation is from Flew's article "Parapsychology Revisited," in *The Humanist*, May 1976.

11. These claims, by the psychic Scientologist Ingo Swann, have been "confirmed" by Harold Puthoff and Russell Targ in experiments reported in chapter 2 of their book *Mind-Reach* (New York: Delacorte Press, 1977).

12. This marvelous technique for winning at roulette by amplifying the precognition of a group of players was explained by Puthoff and Targ in the bound galleys of *Mind-Reach* that were mailed to reviewers by Eleanor Friede of Delacorte Press. After reporting fantastic success with the method they added:

> Such casino exploits have, in fact, stood up to scientific investigation and have resulted in published papers. For those interested, we include here the description of a proven and published strategy. Although somewhat complicated, it has provided a number of individuals known to us an opportunity to succeed at the casino and come away with money in their pockets as testimony to their psychic prowess.

  I commented on this system in a review of *Mind-Reach* reprinted in *Science: Good, Bad and Bogus* (1981). To my surprise, when *Mind-Reach* was published, its pages on the roulette system had vanished. See my book for details.

13. The Levy scandal was reported by Boyce Rensberger in *The New York Times* (August 20, 1974). Other reports appeared in *Time* (August 26, 1974) and the *APA Monitor* (November 1974). Rhine himself discussed the affair (without

mentioning Levy's name) in "A New Case of Experimenter Unreliability," in his *Journal of Parapsychology* (June 1974).

14. In spite of numerous accusations of fraud by C. E. M. Hansel and other skeptics, leading parapsychologists refused to believe the charges until Betty Markwick published her sensational findings in the *Proceedings of the Society for Psychical Research* 56: 250–77, May 1978. For a summary of her evidence, and J. G. Pratt's incredible way of rationalizing it, see *Science: Good, Bad and Bogus*, chapter 19. The Rhine statement about Soal appears in chapter 10 of Rhine's *The Reach of the Mind* (1947).

10. THE FIBONACCI SEQUENCE

1. M. Gardner, *Penrose Tiles to Trapdoor Ciphers* (New York: W. H. Freeman, 1989).
2. A. Horadam, "Fibonacci Number Triples," in *American Mathematical Monthly* 68: 751–53, 1961.
3. V. Schlegel, *Zeitschrift für Mathematik und Physik* 24: 123, 1879.
4. E. B. Escott, *Open Court* 21: 502, 1907.
5. W. Weaver, "Lewis Caroll and a Geometrical Paradox," *American Mathematical Monthly* 45: 234, 1938.
6. T. de Moulidars, *Grande Encyclopédie des Jeux*, p. 459 (Paris, 1888).
7. M. Gardner, *Mathematics, Magic, and Mystery* (New York: Dover, 1956).
8. N. Pandita, *Ganita Kaumudi* (Lotus Delight of Calculation), p. 1356.
9. D. Knuth, *Art of Computer Programming*, vol. 4 (Reading, MA: Addison-Wesley, 2006).
10. M. Feinberg, *Fibonacci Quarterly*, October 1963.
11. O. O'Shea, *The Magic Numbers of the Professor* (Washington, D.C.: Mathematical Association of America, 2007).
12. J. Smoak and T. J. Osler, "A Magic Trick from Fibonacci," *College of Mathematics Journal* 34: 58–60, January 2003.
13. R. Graham, D. Knuth, and O. Patashnik, "Exercise G43," *Concrete Mathematics* (Reading, MA: Addison-Wesley, 1994).
14. Brooke and Wall, *The Fibonacci Quarterly* 1: 80, 1963.
15. M. Gardner, *Mathematical Circus*, (New York: Knopf, 1979).

11. L-TROMINO TILING OF MUTILATED CHESSBOARDS

1. M. Gardner, "The Eight Queens and Chessboard Divisions," in *The Unexpected Hanging and Other Mathematical Diversions* (Chicago: University of Chicago Press, 1991).
2. S. W. Golomb, "Checker boards and polyominoes," *American Mathematical Monthly* 61: 675–82, 1954.

3. S. W. Golomb, *Polyominoes* (New York: Scribner, 1965).
4. R. Nelsen, *Proofs Without Words II: More Exercises in Visual Thinking* (Washington, D.C.: Mathematical Association of America, 2000).
5. I. P. Chu and R. Johnsonbaugh, "Tiling deficient boards with trominoes," *Mathematics Magazine* 59: 34–40, 1986.
6. K. Jones, Vee-21 available at www.gamepuzzles.com/polycub2.htm#V21.
7. C. Jepsen, "On Tiling Deficient Rectangular Boards with Trominoes." *Journal of Recreational Mathematics* 7: 125–30, 1995.

12. IS REUBEN HERSH "OUT THERE"?

1. For instructions on how to make a Klein bottle out of an envelope, see chapter 2, "Klein Bottles and Other Surfaces" in my *Sixth Book of Mathematical Games from Scientific American* (New York: W. H. Freeman, 1971).

19. THE CURIOUS CASE OF FRANK TIPLER

1. Pannenberg was born in 1928 in what now is Poland. His best-known works are *Jesus: God and Man* (1968) and a three-volume *Systematic Theology* (1994), both heavily influenced by Karl Barth. At sixteen he had an experience similar to Paul's on the Road to Damascus. He and Tipler are good friends.

21. WHY I AM NOT AN ATHEIST

1. I suppose it was inevitable that sooner or later someone would apply modern game theory, with its payoff matrices, to the various encounters between God and man in the Old Testament. In any case, Steven J. Brams, professor of politics at New York University, has done it. His book, *Biblical Games: A Strategic Analysis of Stories in the Old Testament*, was published in 1980 by MIT Press, Cambridge, Massachusetts. Chapter 2, "The Meaning of Faith," deals with the two-person strategies involved in the sacrifice stories of Abraham and Jephthah.
2. The sacrifice of Jephthah's daughter has two colorful parallels in Greek mythology: the intended sacrifice of Agamemnon's daughter, and the sacrifice of Idomeneus's son. In one version of the first legend, Agamemnon vows to Artemis that if she provides him with a child he will sacrifice the dearest possession he acquires within a year. The possession is his baby daughter, Iphigenia. Agamemnon understandably refuses to carry out his foolish vow. Years later, on his way to fight the Trojans, his fleet is stranded by lack of wind. A psychic convinces him that Artemis is angry, and that only the sacrifice of Iphigenia will get the ships moving again.
   Agamemnon sends for his daughter, under the pretense of marrying her to Achilles. When she learns the awful truth she consents to the sacrifice, like

Jephthah's daughter, out of her deep "love" for her father, for the gods, and for her country. Just before the knife descends, Artemis takes pity on the girl and spirits her away, leaving a wild deer on the altar. Two surviving plays by Euripedes deal with the legend, along with two operas by Gluck, and plays by Goethe, Racine, and others.

The second legend concerns Idomeneus, king of Crete, who vows to Poseidon that if he gets safely home from Troy he will sacrifice whatever living creature he first encounters. That creature is his son. Idomeneus carries out the murder, but the gods punish Crete with a terrible plague, and his countrymen exile him.

Both legends seem to me less blasphemous than the legend of Jephthah's daughter. Artemis was more merciful than Jehovah, the Cretans more just than the Israelites. It should be said that historical Judaism has strongly denounced Jephthah, and tradition has him punished with a horrible death. In the fifth canto of Dante's *Paradiso*, Beatrice likens Jephthah's crime to Agamemnon's and condemns them both. Similar judgments are in numerous tragedies based on the Jephthah story, and in an opera by Handel. A Victorian painting by Sir John Millais shows Jephthah in deep sorrow, being comforted by his simpleminded daughter. I find the picture as preposterous as Byron's poem "Jephthah's Daughter," which ends with the girl saying to her father, "And forget not I smiled as I died!"

One marvels at the mind-set of those Christian commentators who praised Jephthah's "faith," as though somehow his insane trust in Jehovah was more admirable than the trust of Agamemnon and Idomeneus in their gods. Some commentators even regarded Jephthah and his daughter, like Abraham and Isaac, as foreshadowings of the Atonement, with special reference to the prayer of Jesus, anticipating his execution, "Not my will but thine be done."

3. Joseph Butler, in his *Analogy of Religion*, writes:

> Indeed a proof, even a demonstrative one, of a future life, would not be a proof of religion. For that we are to live hereafter is just as reconcilable with the scheme of atheism, and as well to be accounted for by it, as that we are now alive is; and therefore nothing can be more absurd than to argue from that scheme, that there can be no future state.

4. The well-known passage in which Kant speaks of the two things that most fill him with awe, the starry heavens above and the moral law within, came to mind when I encountered in the second volume of Bertrand Russell's autobiography a paragraph that Kant would surely have found mystifying. Russell was near finishing his *Practice and Theory of Bolshevism* (1920), and he is trying to decide if he should publish his book:

> To say anything against Bolshevism was, of course, to play into the hands of reaction, and most of my friends took the view that one ought not to say what

one thought about Russia unless what one thought was favourable. I had, however, been impervious to similar arguments from patriots during the War, and it seemed to me that in the long run no good purpose would be served by holding one's tongue. The matter was, of course, much complicated for me by the question of my personal relations with Dora. One hot summer night, after she had gone to sleep, I got up and sat on the balcony of our room and contemplated the stars. I tried to see the question without the heat of party passion and imagined myself holding a conversation with Cassiopeia. It seemed to me that I should be more in harmony with the stars if I published what I thought about Bolshevism than if I did not. So I went on with the work and finished the book on the night before we started for Marseilles.

5. Miguel de Unamuno, *The Tragic Sense of Life*, chapter 8.
6. The basic argument of Pascal's wager seems to have been first explicitly given by Arnobius, a Christian theologian who lived about A.D. 300 in Africa:

> But Christ himself does not prove what he promises. It is true. For, as I have said, there cannot be any absolute proof of future events. Therefore since it is a condition of future events that they cannot be grasped or comprehended by any efforts of anticipation, is it not more reasonable, out of two alternatives that are uncertain and that are hanging in doubtful expectation, to give credence to the one that gives some hope rather than to the one that offers none at all? For in the former case there is no danger if, as is said to threaten, it becomes empty and void; while in the latter case the danger is greatest, that is, the loss of salvation, if when the time comes it is found that it was not a falsehood.

I quote from Augustus De Morgan's *Budget of Paradoxes*, Vol. 2 (1872), where a footnote supplies a translation of the Latin passage which De Morgan quotes from Pierre Bayle, a seventeenth-century French Protestant of extreme fideist views. "Really Arnobius seems to have got as much out of the notion," comments De Morgan, ". . . as if he had been fourteen centuries later, with the arithmetic of chances to help him."
7. Count Manuel's remarks are almost a paraphrase of the following passage from Cicero's essay "On Old Age":

> And after all, should this my firm persuasion of the soul's immortality, prove to be a mere delusion; it is at least a pleasing delusion; and I will cherish it to my latest breath. I have the satisfaction in the meantime to be assured, that if death should utterly extinguish my existence, as some minute philosophers assert; the groundless hope I entertained of an after-life in some better state, cannot expose me to the derision of these wonderful sages, when they and I shall be no more.

8. George Santayana, *Character and Opinion in the United States* (1920).

9. Raymond Smullyan, mathematician and magician, has this to say about science and magic in a whimsical essay on astrology in his book *The Tao Is Silent* (1977):

> Speaking of magic, I am genuinely open to the possibility that the entire Universe works ultimately by magic rather than by scientific principle. Who knows, perhaps the Universe is a great magician who does not want us to suspect his magical powers and so arranges most of the visible phenomena in a scientific and orderly fashion in order to fool us and prevent us from knowing him as he really is! Yes, this is a genuine possibility, and the more I think about it, the more I like the idea!

Smullyan's suggestion is, of course, not far from the Hindu doctrine that the entire universe is an illusion. Although the medieval Schoolmen thought the universe was real enough, many of them (notably William of Occam) argued that God was quite capable of annihilating, say, a chair or a star, substituting for it the illusion of a chair or a star which we would be unable to distinguish from a real one.

10. Miguel de Unamuno, *The Tragic Sense of Life*, chapter 9.

INDEX

Page numbers in *italics* refer to figures.

immortality, 191, 193–94, 196, 197, 198, 199–200, 202–203, 204, 205, 217–18, 231*n*
Incarnation, 185
Ingersoll, Robert G., 193
intelligent design (ID), 3, 5–8, 20–21, 23
International New Thought Alliance (INTA), 50, 51, 57*n*, 69
Intimate Letters of Archie Butt, The (Butt), 89
Iran, 225–26
Irving, Washington, 147, 148
Isaac Asimov's Science Fiction Magazine, 14, 103, 131, 134
"Isaac Newton: Alchemist and Fundamentalist" (Gardner), 9

James, William, 41, 190, 191, 200–203, 204, 209
Jaroff, Leon, 47
Jensen, Christopher, 122
Jephthah, 192, 230*n*–31*n*
"Jephthah's Daughter" (Byron), 231*n*
Jesus, 7, 10, 24, 32, 38, 39, 51, 55, 64, 69–70, 146, 154, 183–84, 186, 187, 188, 191, 193, 226, 231*n*, 232*n*
Jesus: God and Man (Pannenberg), 230*n*
Jewels of Memory (Joyce), 60
Jews, Judaism, 182, 185, 192, 213, 231*n*
Jinn from Hyperspace (Gardner), 124
Job, Book of, 212, 219
John Paul II, Pope, 22
Johnson, Phillip, 20, 21
Johnson, Samuel, 18, 40, 218
Johnsonbaugh, Richard, 121
John W. Luce and Company, 169
Jones, Kate, 122
Jorkens, 154
Joseph, Mary and, 183
Joshua, 32, 182, 215
Journal of Magnetism, 57*n*
Journal of Recreational Mathematics, 106
Journal of the American Society for Psychical Research, 87, 90, 92, 93
Joyce, John Alexander, 60–61
Jurist, Janet, 176

Kadon Enterprises, 122
Kant, Immanuel, 190, 193, 197–99, 204, 218, 231*n*
karma, 64, 69, 87
Kepler, Johannes, 18, 37
Keynes, John Maynard, 10
"Kidnapped Santa Claus, A" (Baum), 152
Kierkegaard, Søren, 192, 194, 201
Kipling, Rudyard, 164–65, 169, 175
Klein bottles, 126, 230*n*
Knuth, Donald E., 111, 113
Koestler, Arthur, 84
Kriss Kringl (Christkindl), 146
Kulagina, Nina, 45
Kung, Hans, 7, 22
Kushner, Harold, 214

Lamarck, Jean-Baptiste, 24–25
Larned, Linda Hull, 150
Late George Apley, The (Marquand), 38
Laughing Dragon of Oz, The (Baum), 142–43
laws, scientific, 30, 31–32
leap of faith, 194–95, 201–202, 203, 204, 206
Lehmann, Ingmar, 106
Leibniz, Gottfried Wilhelm, 12, 126, 216–17
Leslie, Frank, 58
Lessons in Truth (Cady), 68
"Let Platonism Die" (Davies), 124
levitation, 32–33, 34, 38
Levy, Walter J., Jr., 47, 228*n*–29*n*
Lewis, C. S., 218
liberals, liberalism, 3–8
Life and Adventures of Santa Claus, The (Baum), 145, 149–52, 175
light, 11
Light, 89
Lincoln, Abraham, 61
live options, 200–202, 203, 204
Locke, John, 12
Locksley Hall (Tennyson), 18
"Locus of Mathematical Reality, The" (White), 127

Printed in the USA
CPSIA information can be obtained
at www.ICGtesting.com
LVHW091141150724
785511LV00005B/456

9 780374 532413